HISTORIC
ILLINOIS
FROM THE AIR

HISTORIC
ILLINOIS
FROM THE AIR

D A V I D   B U I S S E R E T

Illustrations and Cartography by Tom Willcockson

THE UNIVERSITY OF CHICAGO PRESS

Chicago and London

DAVID BUISSERET is the director of the
Hermon Dunlap Smith Center for the History of
Cartography at the Newberry Library.
He is the editor of *From Sea Charts to Satellite Images:
Interpreting North American History through Maps*,
recently published by the University of Chicago
Press, and author of *Historic Jamaica from the Air*.

The University of Chicago Press, Chicago 60637
The University of Chicago Press, Ltd., London

© 1990 by The University of Chicago
All rights reserved. Published 1990
Printed in the United States of America

99 98 97 96 95 94 93 92 91 90 54321

Library of Congress Cataloging-in-Publication Data

Buisseret, David.
    Historic Illinois from the air/ David Buisseret : illustrations
and cartography by Tom Willcockson.
        p.   cm.
    Includes bibliographical references.
    ISBN 0-226-07989-9 (cloth)
    1. Illinois—Aerial photographs.   2. Illinois—Description and
travel—1981– —Views.   3. Historic sites—Illinois—Pictorial
works.   I. Title.
F542.B85   1990
917.73'0022'2—dc20   89-20648
                                                        CIP

This book is printed on acid-free paper.

PHOTO CREDITS

Tim Buisseret, figs. 175, 287
Richard Norrish, fig. 16
Dave Pasquith, fig. 115
Ernst Seinwill, figs. 6, 7, 11, 44, 46, 51, 57, 65, 77, 82, 121,
123, 124, 127, 145, 162, 176, 192, 205, 215, 217, 219, 224,
229, 231, 238, 241, 243, 247, 252, 258, 272, 294, 297
Tom Willcockson, figs. 144, 155

For Pat and Terry

CONTENTS

7 · CHICAGO AND THE EXPANSION OF THE NORTHEAST

8 · RAILROADS IN ILLINOIS

9 · CHICAGO BECOMES THE SECOND CITY

10 · THE MATURE CITY AND ITS INSTITUTIONS

11 · THE STATE COMES OF AGE

ILLUSTRATIONS

INTRODUCTION

I REMEMBER WELL the day in 1955 when the possibilities of aerial photography first became plain to me. After two years of involuntary service with the British Army in Egypt, we were being flown back to England in an Avro York, which was a summarily converted Lancaster bomber. First we flew over the Nile Delta, with its astonishingly sharp contrast between the irrigated area and the desert, and then we worked along the coast westwards, towards Tobruk.

Soon one of the four engines began to fail, and we flew lower and lower, with a remarkable view of the field systems along the North African coast. I still do not know if they were the work of the ancient Romans or of the then-recent Italians, but the point was that when you could see *nothing* on the ground—which we eventually reached at El Adem—you could from the air see a whole series of land-cultivation patterns.

At the time I was unaware that aerial archeology had cut its teeth about forty years earlier on similar views of the Syrian desert, but I did realize that this was a field with fascinating possibilities for the historian. Returning to England, I went to Cambridge University as an undergraduate and joined the University Air Squadron. Much of our flying was carried out more or less upside down, which offered remarkable views of the English countryside with its innumerable historic features stretching over at least two millennia: early dikes, hill forts, hut circles, eighteenth-century planned towns, Victorian forts, and so forth.

For a good many years I was unable to turn this private interest into a professional pursuit. Naturally there was no undergraduate course at Cambridge involving aerial photography for historians, and my eventual doctoral-thesis subject concerned seventeenth-century French finances, an area where aerial photography is not very useful. In 1964 I went to teach history at the University of the West Indies on their Jamaica campus. The flying there was marvellous, with usually clear weather, an absence of other aircraft, and stunning views of the sea and mountains.

The possibilities for using aerial photography to investigate the human past were limited, since the tropical vegetation tends to obscure buildings and even field patterns quickly and thoroughly. Still, in 1968 Jack Tyndale-Biscoe and I did publish *Historic Jamaica from the Air*, which largely relied on comparisons between aerial photographs and historic maps to elucidate various features now obscured.

Moving to The Newberry Library in Chicago in 1980, I seemed at first to be going to another region with little obvious potential for historical investigation through aerial photography. However, Richard Brown, director of research and education at the library, encouraged me to think of applying this technique to Illinois history, and as time went by it became plain that there was indeed great potential here. To start with, Chicago had been since the 1920s one of the great centers for commercial aerial photography in the United States, and material from companies like the Chicago Aerial Survey could be found in various collections. These were mostly low-level, oblique, black-and-white photographs, and to complement them there were fine collections of medium-level, vertical, black-and-white photographs available at the map department of the library of the University of Illinois at Champaign-Urbana, where they had been deposited by various federal agencies. Similar photographs were to be found at the Illinois Department of Transportation (Springfield), which had by chance sometimes taken photographs of interest to historians, in the course of work on roads and bridges.

The most difficult material to obtain was good-quality satellite imagery, but in 1985 the Illinois

State Geological Survey published images of parts of Illinois at different scales, so that problem was solved. There remained the question of how to obtain low-level, oblique photographs of areas of interest not already covered. Here I was fortunate to have at first the services of Ernst Seinwill, who had once been a photographer with the Chicago Aerial Survey and who worked freelance with me on several flights. When Ernst retired to New Mexico in 1986, I took advantage of a generous offer from Lewis University to use the services of their flying school and have been able myself to complete most of what was needed, with the help of pilots Tim Drake and Ole Doyle.

For all these local photographs we used light aircraft, shooting out of an open side window. This is not the ideal procedure because, particularly in high winds, it is not easy to attain exactly the right attitude of the aircraft so as to avoid intruding wings and struts. Nor was our photographic equipment ideal, since for many images we had to rely on enlargements from 35-mm film, whereas a larger size would have been more desirable for greater detail. We have to remind our readers that this is primarily a particular kind of history book, rather than a demonstration of photographic excellence.

In writing the text I was influenced by the ideas of Raymond Chevallier in France (*La photographie aérienne*, Paris, 1971), W. G. Hoskins in England (particularly *English Landscapes*, London, 1974), and J. B. Jackson of the United States. Jackson set out in the inaugural issue of his review, *Landscape*, a description of the potential of aerial photography that could hardly be bettered.

It is from the air that the true relationship between the natural and the human landscape is first clearly revealed. . . . No one who has experienced this spectacle . . . can have failed to be fascinated by it, nor wonder at the variety of man's ways of coming to terms with nature. Why are some stretches of land thickly settled with villages almost within sight of one another, while others are occupied by great rectangular fields and a few lonely homes? . . . And to such questions an equally varied set of answers occurs . . . [but] the asking of such questions is more important than the finding of an answer. It means that, like the air traveller, we have acquired a new and valuable perspective on the world of men, and that with it eventually comes the realization there really is no such thing as a dull landscape or farm or town.

Curiously, *Landscape* made little use of aerial photography in carrying out this part of Jackson's program. But his words accurately state the aims of this book. Readers should note, however, that we are concerned here only with those aspects of Illinois history that can profitably be illustrated and interpreted from the air; we have nothing to say, for instance, about voting patterns or developments in political thought and do not stray much into the twentieth century. The bibliographies for each chapter are similarly limited to those books that we found most useful for our particular purpose.

Other limitations arose from the nature of the subject. Sometimes we were able to procure a good image of a certain site when we could find or take only an inferior image of a better or more typical site; in such cases we preferred to go with the better image. The reader may well complain that we ought to have gone on looking and flying until we had perfect images for all the most suitable sites. This would not have been practical, given our tight constraints on time and money, which always dictated some compromise in order to get the job done.

Many people helped with the project. In the Newberry Library, John Long was an early collaborator and has remained a helpful critic, as have Jim Akerman, John Aubrey, and Robert Karrow. At the University of Illinois at Champaign, I received early encouragement and advice from John Hoffmann and Robert Sutton, and much help in the map room from David Cobb. Among the universities in Chicago, I have been helped by Michael Conzen of the University of Chicago, by Gerald Danzer of the University of Illinois at Chicago, and by Theodore Karamanski of Loyola University.

When the book was more or less planned and many of the photographs had been taken, I had the good fortune to fall in with Tom Willcockson, who joined the project in order to draw explanatory maps and to compose aerial perspectives. In the event, he did much more than that, participating in the planning (and expansion) of all the chapters and in their research. It may seem odd that two people raised outside Illinois should want to write a book about the state, but we believe that just because we do come from outside, we find Illinois much more exciting as a place for historical investigation than do many people born and bred here. In a book touching on so many places and themes, we are sure as outsiders and neophytes to have made mistakes. But we shall be content if, as Jackson put it, we have offered for some people "a new and valuable perspective on the world of men," persuading others that "there really is no such thing as a dull landscape or farm or town."

I

THE LAND

1. LANDSAT image of Illinois, September–
October 1982 (Illinois State Geological Survey and
Northern Illinois University)

Introduction: General Topography

THE STRIKING IMAGE of Illinois in figure 1 was derived from information collected by a LAND-SAT satellite during the fall of 1982. It combines information from seven separate parts of the spectrum, ranging from visible light to thermal infrared emissions, and is of course in false color. As the fields were bare after harvest, they were emitting little heat and so show up as white areas; water is dark blue, built-up areas violet, and vegetation red. The satellite image commentates in a remarkable

2. Map showing main types of land surface

A) Driftless Region
   (Unglaciated)
B) Eroded Glacial Drift
   (Illinois Glacier)
C) Recent Glaciation
   (Wisconsin Glacier)
D) Former Lake-bed
   (Post-Wisconsin Glacier)

way the landform pattern of the state as it has emerged over the millennia.

Figure 2 shows the major types of land surface that can be identified in the image. In the far south of the state and in the extreme northwest are two driftless areas (the Shawnee Hills and the Wisconsin driftless area), so called because they never underwent glaciation; these lands are consequently very hilly and heavily wooded, which gives them the dark red signature. From the Mount Vernon hill country northwest to the Rock River hill country is a great crescent of gently rolling land, punctuated by numerous rivers and streams; this area marks the outer reaches of the Illinois glacier. After the ice withdrew, this land began to be eroded by drainage systems, which appear as red fingerlike forms, the mark of the heavy vegetation along these new river valleys.

A large block in the northeastern part of the state (the Bloomington plain) is the region of most recent glaciation, from which the Wisconsin glacier retreated in relatively recent times. Here the land is flat, with low moraines, and drainage systems have not yet fully established themselves. The soil is deep and when drained and tilled makes excellent agricultural land; in the satellite image this area is divided into large geometric fields, growing crops such as corn and soybeans.

Other distinctive areas in figure 1 are the alluvial floodplain of the Mississippi River, running along the western boundary of the state, and the former Chicago lake bed, now largely covered with buildings which give a bluish-grey signature. The ancient runoff from Lake Michigan into the Illinois River can just be seen to the west of the city; in the 1840s the Illinois and Michigan Canal reestablished this link and began the extraordinary growth of Chicago (chapter 6). Apart from the geometric patterns of the farmland, whose emergence is discussed in chapter 4, the major man-made features visible in figure 1 are the airports in the north of the state and the reservoirs in the south. But these are dwarfed by the pattern of the natural landforms, which have always controlled the development of the state.

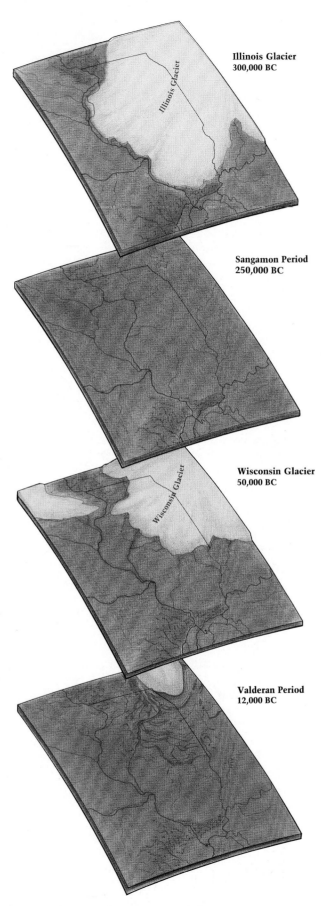

Illinois Glacier
300,000 BC

Sangamon Period
250,000 BC

Wisconsin Glacier
50,000 BC

Valderan Period
12,000 BC

3. Maps showing the sequence of glaciers

The Work of Ice and Water

THE MOST IMPORTANT INFLUENCE on Illinois's landforms was the advance and slow retreat of four glacial systems. The last two of these, whose influence is most obvious, were the Illinois glacier and the Wisconsin glacier, the latter lasting from fifty to fifteen thousand years ago. The first map in figure 3 shows the area of future Illinois at the maximum extent of the Illinois glacier. The whole state was covered, except for a patch in the extreme northwest and in the far south. These patches were not subject to the relentless levelling and grinding of the great sheet of ice and so survive to this day, as we saw on the satellite image, as hilly, heavily wooded regions.

As the glacier withdrew, it deposited huge quantities of minerals on the land and sent out huge quantities of meltwater, which drained into the Gulf of Mexico. At that time the main north-south stream flowed in the bed of what is now the Illinois River. The glacier eventually retreated, leaving the whole central area ground down and covered with mineral deposits.

The next (and so far final) glacier was the Wisconsin, which at its furthest extent reached about halfway down the state. As it advanced, it pushed the main north-south river out of the bed of the Illinois River and into what we know as the Mississippi River. It also ground deeper into what became the lake floor of Lake Michigan, and it eventually filled the Illinois River valley with rich deposits. In its retreat it left further deposits (moraines) on the upper half of the state (always excluding, of course, the driftless area of the northwest). The meltwater formed a huge Lake Michigan, which for a time drained into the Illinois River valley, carving through the recently deposited matter and leaving the great bluffs at sites like Starved Rock. Even in terms of human history, the departure of the Wisconsin glacier was not very long ago, and as we see from figure 1, its influence is still very much with us.

Figure 5 shows how the future area of Chicago was affected in the last stage of glaciation, as the icy waters from newly created Lake Michigan sought an exit through a great dam of morainic material and down the Illinois River valley. Various knobs of relatively hard material in the moraines stood up to the passage of the waters, and as time went by and the

4. Aerial photograph showing successive beachlines at Illinois Beach State Park (no date or provenance known; Newberry Library)

level of the lake receded, these knobs became areas of high land such as Blue Island and Stony Island. At a later stage, the north-south lake current deposited successive layers of sand along the beach, until it reached the gentle curve that we know today. In an aerial photograph like figure 4 this beach-building process is graphically demonstrated; the striations in the sand show that the beachlines, which began by tending quite strongly to the west, eventually ran almost due north-south, as today's shore was created.

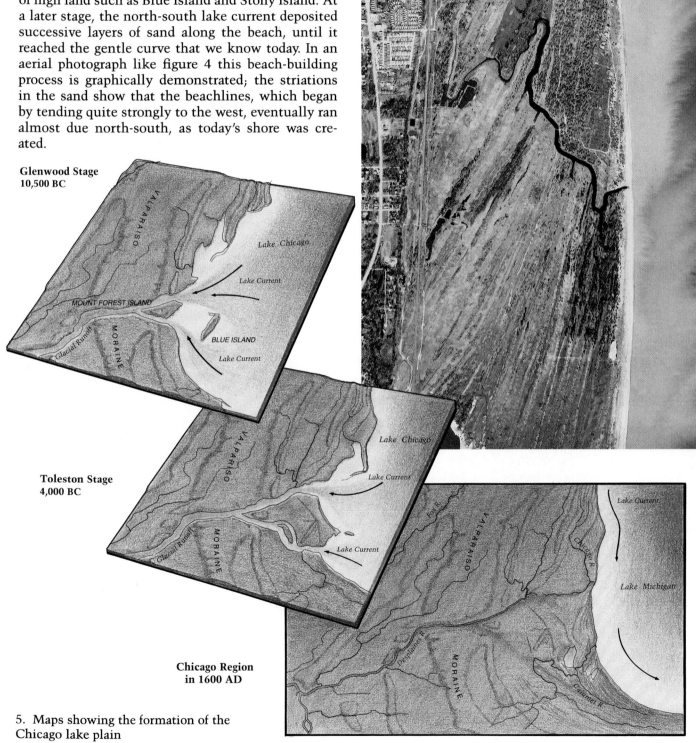

Glenwood Stage
10,500 BC

VALPARAISO
Lake Chicago
Lake Current
MOUNT FOREST ISLAND
Glacial Runoff
MORAINE
BLUE ISLAND
Lake Current

Toleston Stage
4,000 BC

VALPARAISO
Lake Chicago
Lake Current
Glacial Runoff
MORAINE
Lake Current

Chicago Region
in 1600 AD

Fox R.
VALPARAISO
Chicago R.
Lake Current
Desplaines R.
MORAINE
Lake Michigan
Calumet R.

5. Maps showing the formation of the Chicago lake plain

River Systems

ILLINOIS IS MOSTLY BOUNDED by waterways. The Mississippi defines the state in the west, the Ohio and the Wabash border it in the southeast, while Lake Michigan marks the northeast corner. Only in the north and northeast does it have the straight, geometrical border characteristic of most midwestern states. The rivers have not been political barriers, like the rivers hemming in so many of the powers of Europe, but at some periods they were economic barriers, for instance determining that opposite banks of the Mississippi would develop differently during part of the eighteenth century.

The rivers along Illinois's borders are varied in character. To the west, the Mississippi and Missouri draw their water out of the immense heart of the country, carrying enormous quantities of sediment. To the east the Ohio, which has a shorter and rockier course, is relatively clear, as may be seen in figure 6, showing the confluence of the Mississippi (left) and the Ohio (right) at Cairo.

The silt-laden Mississippi played a large part in soil formation, for all along its bank from Saint Louis to Cairo were deposited loams, some of which

6. Aerial photograph of the confluence of the Mississippi and Ohio rivers, September 1986

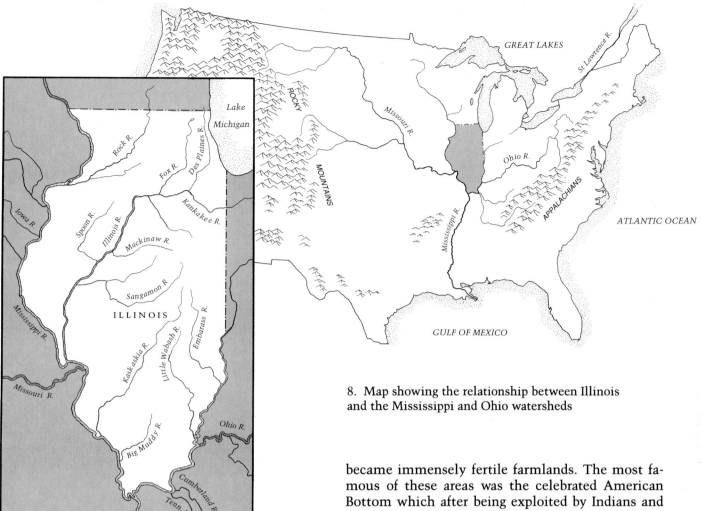

8. Map showing the relationship between Illinois and the Mississippi and Ohio watersheds

7. Aerial photograph of a meander pattern on the Wabash River, September 1986

became immensely fertile farmlands. The most famous of these areas was the celebrated American Bottom which after being exploited by Indians and French became for a time the heart of midwestern agriculture. Many other areas of spectacular fertility lay alongside the rivers and to a considerable degree dictated the early pattern of European settlement.

The rivers also influenced the routes of communication from the time of the Indians right up to the present, whether they bore canoes, steamboats, or huge barge tows like those in figure 6. They were by no means constant in their courses, however, and this could sometimes lead to disastrous floods in towns like Cairo. Even the peaceful Wabash River has often changed its bed, outlining swales among the fields on its banks (figure 7); when the Mississippi River decides to change course, the results can be dramatic indeed.

As figure 8 shows, Illinois sits astride a vast network of water communications, reaching from Minneapolis in the north to the Rocky Mountains in the west. To the south the Mississippi leads to New Orleans and the gulf, and in the east the Ohio provides access all the way to the Appalachians. Lake Michigan and the Great Lakes extend the reach of Illinois from Duluth on Lake Superior to the Saint Lawrence and the Atlantic. Before the advent of the railroads, these water communications were crucial, and they remain very important even today.

9. Map showing forests and prairies about 1800

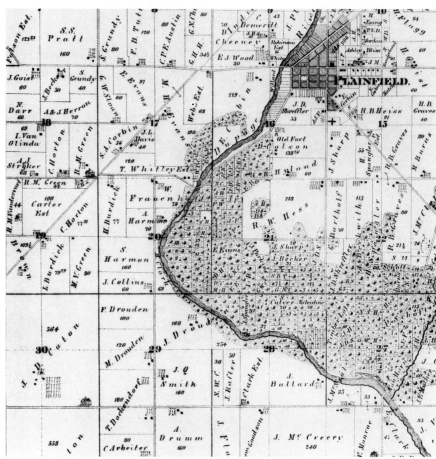

10. Detail of the Plainfield area (*Atlas of Will County*, Chicago, 1873; Newberry Library)

Forests and Prairies

U P TO THE BEGINNING of the nineteenth century, the Europeans settling North America had contended almost exclusively with heavily forested country, for which they had developed appropriate technologies. In the future state of Illinois, however, only about half the land was forest about 1800, as figure 9 shows; the rest was under the tall grasses known by the French word *prairie*, though in truth these prairies did not much resemble the green and short-cropped meadows of northern France after which they were named.

Because the prairie in Illinois eventually became an immensely productive agricultural region, heavily settled, there has been much scholarly debate about why the woodlands were unable to take hold in so large an area of the state. Looking at figure 9, we are immediately struck by the coincidence between areas of woodland and relatively hilly regions shown in figure 2; clearly this broken country was

favorable for the growth of trees, as were the river bottoms. But why did the open areas—corresponding to the area covered by the Wisconsin glacier (figure 3)—largely remain treeless?

Some scholars have dwelt on strictly climatic factors, such as the levels of rain- and snowfall, strength of the winds, mean average temperatures, and so forth. These, along with the landforms—areas of slow drainage favored grasses, whereas broken country was more suitable for trees—clearly played a part. But the evidence of many of the early maps suggests that scholars of another school have identified an important element.

According to this theory, trees failed to colonize the open spaces because these were ravaged by periodic fires, whether started by lightning or by Indians hunting game. While the deep-rooted grasses could survive a fire and even thrive after it, saplings tended to be burnt out. Some accounts of early prairie fires are truly terrifying, and we have tried to give a faint idea of what they must have looked like in figure 11, showing a stubble burning near Chester.

Evidence from the northern part of the state

11. Aerial photograph of a field fire in southern Illinois, September 1986

seems to show that these fires, usually fanned by breezes from the southwest, raged unimpeded over the open prairie until they were stopped by an obstacle such as a roughly north-south river. Figure 10 shows one such river near Plainfield. Here the Du Page River runs roughly north-south, and the area to the west of it is bare of trees, whereas in the sheltered bend of the river they are thriving. It is true that this example dates from about 1870, but the same phenomenon could be shown repeatedly on the General Land Office maps of the early nineteenth century, drawn when Europeans first reached this area. Of course, the fire theory does not explain phenomena such as the periodic incidence of "groves" of trees on the open land, but it does seem to have been one important variable among the many that determined the distribution of vegetation in the state.

12. Aerial photograph of geese flying up the Illinois River, April 1987

Illinois Wildlife

THE GREAT VARIETY OF LANDFORMS and vegetation in Illinois gave rise to widely varied populations of birds and animals. In the northeast, fish-eating birds inhabited the dunes and swamps by Lake Michigan. The unglaciated, forested areas of the far south and the far northwest sheltered a variety of animals, including bears and large predators such as wildcats. The prairie had its own ecosystems, with huge herds of bison and unusual animals such as the prairie dogs as part of its food chain. The areas where prairie and forest met were peculiarly suitable for deer, elk, and wolves. By the Mississippi River was a great flyway seasonally filled with ducks and geese, where eagles could also be seen, feeding on the abundant supply of fish.

Most of these birds and animals suffered from the European intrusion, some to the point of extinction. Lake Michigan was overfished and then polluted by a variety of wastes, so that the early accounts of abundant fish now seem like travellers' fantasies. The tall prairie grasses gave way to cultivated fields, lacking both food and cover for the many animals

that had made the prairies their home. In the timbered part of the flat land, the woods by the rivers were often the first place to be occupied, with a corresponding decline in the wildlife. Eventually even the Mississippi itself came to be more or less polluted, a process that is beginning to be reversed in our time.

Not all birds and animals suffered from these changes. Some, like the deer, even profited from the new landscape, in which field alternated with forest. In the twentieth century deer began to proliferate in embarrassing numbers, occupying such "new" terrain as the woodland inside O'Hare International Airport. Small animals such as raccoons, skunks, and squirrels were also able to find their niches in the new ecosystems, often becoming satisfied suburbanites. Birds, too, have shown a remarkable resilience, some returning to areas from which high levels of pesticide contamination had driven them, and

13. Aerial photograph of deer in a snow-covered forest about 1950 (Illinois Natural History Survey)

others, like the Canada geese, taking full advantage of the new conditions—in their case, the patches of water newly free of ice throughout the winter.

Sometimes this wildlife looks spectacular from the air. Figure 12 shows a small flock of geese making its way northwards up the Illinois River in the spring, and figure 13 shows deer in a snowy wood in Ogle County. Winter, indeed, has proved to be a good time to count large-animal populations from the air. Whereas in areas like the Serengeti plain of Africa aerial spotters have to count animals against their natural camouflage, in the winter in Illinois large animals like deer stand out sharply from a background of three or four inches of snow. Naturalists can make accurate counts of deer and in some cases can recommend action to control overgrazing.

2

THE AMERINDIANS

Introduction: The Indians of Illinois

THE HISTORY OF THE AMERINDIANS who occupied Illinois before the advent of the Europeans is complex and still not well understood. We do know that they were descendants of the nomadic hunters who crossed from Asia before 25,000 B.C., and that by 8000 B.C. a group of them lived near Modoc (figure 15). By about 1000 B.C. these peoples, or their successors, were entering what is called the "woodland phase," characterized by limited agriculture and the formation of centers of population.

14. George Catlin, portrait of a Peoria chief (National Museum of American Art, Smithsonian Institution, Gift of Mrs. Joseph Harrison, Jr.)

From 1000 A.D. onwards, we see the beginnings of the "Mississippian culture," which thrived on the fertile floodplains of the Mississippi and especially around Cahokia. By the time of Columbus this culture had developed an extensive agriculture and trade network and had built remarkable mounds, some of which have survived.

About 1500, the story goes, this thriving culture suffered some catastrophic, mysterious decline, so that the first Europeans to reach this area dealt with quite different groups under such vaguely familiar names as Sauk, Fox, Potawatomi, and Illini. These people also left earthworks and were part of a widespread trade network.

The later Indians belonged to tribes that were often on the move, sometimes from winter camps to summer planting and hunting grounds and often in response to threats from powerful neighbors like the Iroquois tribes, raiding from the east. Figure 28 shows roughly where these tribes were in the seventeenth and eighteenth centuries. Europeans wrote descriptions of these later Indians, who left many artifacts, recovered archeologically or handed down to living successors. The later mounds, some of which survive, often represented effigies of birds and animals that were important as prey or sacred creatures—or both.

The general impression is that these later Indians were more or less effaced from the state in the early nineteenth century, as a result of treaties with the federal government that pushed them across the Mississippi River. Recent research suggests, however, that many of them did not observe the provisions of the various treaties, but hung on in obscure corners of the thinly settled state and eventually were absorbed into the larger society. Hence, perhaps, the many Illinois families that boast of having "Indian blood"; their stories may be more often founded in fact than was once thought. The tribes also live on in an unexpectedly large number of Illinois place-names.

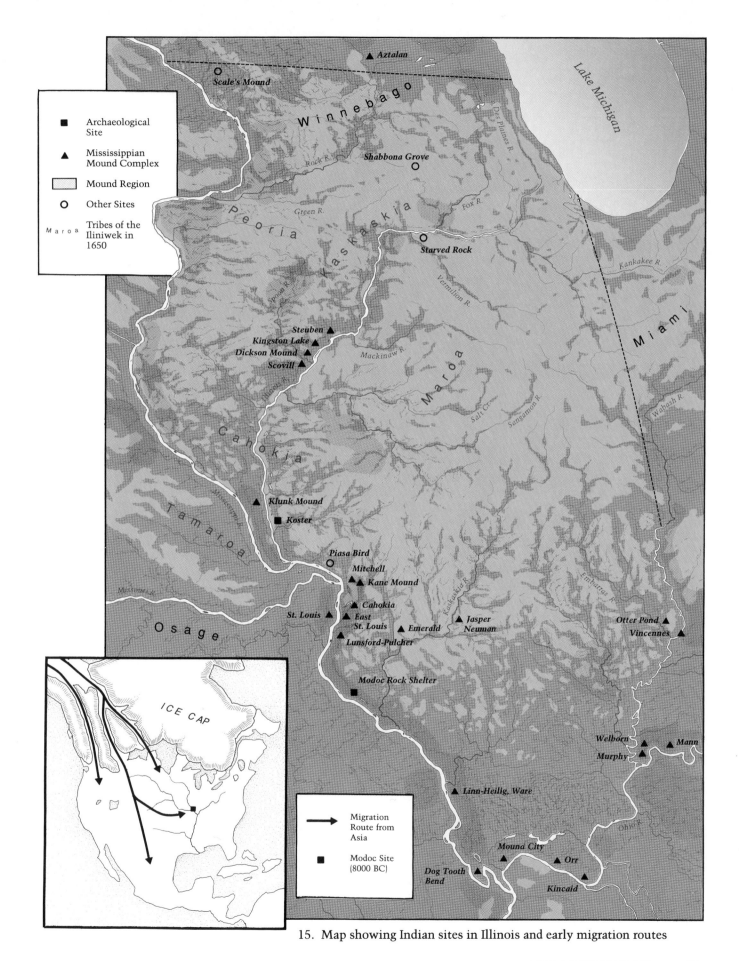

Legend

- ■ Archaeological Site
- ▲ Mississippian Mound Complex
- ▒ Mound Region
- ○ Other Sites
- *Maroa* Tribes of the Iliniwek in 1650

Aztalan

Scale's Mound

Winnebago

Lake Michigan

Des Plaines R.

Rock R.

Shabbona Grove

Peoria

Green R.

Fox R.

Kaskaskia

Starved Rock

Kankakee R.

Vermilion R.

Spoon R.

Steuben
Kingston Lake
Dickson Mound
Scovill

Mackinaw R.

Maroa

Miami

Illinois R.

Salt Cr.

Sangamon R.

Wabash R.

Cahokia

Mississippi R.

Klunk Mound
Koster

Tamaroa

Missouri R.

Piasa Bird
Mitchell
Kane Mound
Cahokia
St. Louis
East St. Louis
Emerald
Jasper Neuman
Lunsford-Pulcher

Kaskaskia R.

Embarras R.

Otter Pond
Vincennes

Osage

Modoc Rock Shelter

Welborn
Murphy
Mann

Linn-Heilig, Ware

ICE CAP

- → Migration Route from Asia
- ■ Modoc Site (8000 BC)

Mound City
Dog Tooth Bend
Orr
Kincaid

Ohio R.

15. Map showing Indian sites in Illinois and early migration routes

16. Aerial photograph of Monks Mound, spring 1983 (Richard Norrish, Edwardsville)

The Early Societies of the Mississippi Valley

EARLY IN THE 1800s Europeans began to write about some fantastic ruins near the confluence of the Mississippi and Missouri rivers, theorizing that these works might have been constructed by ancient Hindus. The most extensive of these ruins were at Cahokia, and we now know that this was one of the thriving centers of the Mississippian culture and in about 1250 may have had a population of about ten thousand.

Figures 16 and 18 give us a good idea of the central area of what is one of the most sensational Indian sites in North America. Monks Mound is named for some French Trappist monks who lived nearby in the early nineteenth century. It rises in four terraces to a height of one hundred feet, where the platform probably supported a structure used as a residence and for ceremonies. This is the largest prehistoric earthern construction in North America and with its well-defined terraces is reminiscent of the ceremonial masonry mounds of Central America.

Figure 18 shows the site as it might have been about 1500. Monks Mound is in the background, and scattered over a roughly lozenge-shaped area about three miles long by two miles wide are many other mounds, some conical, some ridge-topped, and some flat-topped. These would have been used as residences for chiefs, as burial sites, and for ceremonies. At ground level lay broad plazas and hundreds of houses, all of which were surrounded by an elaborate stockade fence.

The Mississippian culture seems to have developed from about 900 A.D. onwards, taking advantage of the rich soil of the Mississippi floodplain. In time it developed an extensive trade network, so that Cahokia artisans worked with Florida seashells, Lake Superior copper, Wyoming obsidian, and Carolina

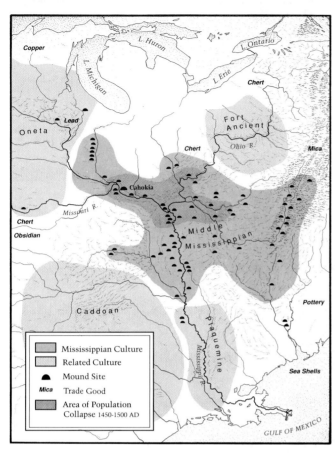

17. Map showing location of the Mississippian and related cultures

mica. The characteristic mounds required an enormous amount of labor and argue for a relatively centralized form of government. This culture also had some knowledge of astronomy. Three thousand feet west of Monks Mound are five circular patterns formed by twenty-four-inch-diameter posts set about four feet into the ground. These posts, which may have been as much as thirty feet high, formed part of a system for marking solstices and equinoxes; because of their affinity with the British astronomical site of Stonehenge, the circles are known as Woodhenge.

One of the greatest mysteries about the Cahokia site is the fate of the peoples who built it. Some scholars believe that they fell prey to European diseases, brought up the Mississippi River. But others argue that the Cahokians might have outgrown their resources or fallen into a period of social unrest and disintegration. Whatever happened to them, their great site was rapidly forgotten, so that the French explorers Marquette and Joliet could in the 1670s pass close by it without noticing its existence. Many of the Mississippian mounds have been destroyed, either by erosion or by some form of development, but the Cahokia site is now protected by the Illinois Department of Conservation. Future investigations may give us a better idea about the peoples of Cahokia and about their relationship to the Indians who followed them.

18. Aerial perspective of the Cahokia Mounds site at its greatest extent

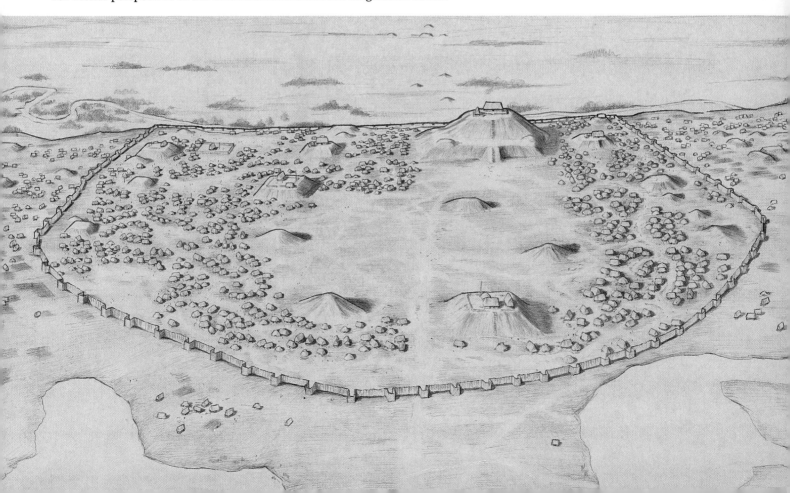

19. Aerial photograph of the country near Scales Mound, April 1988

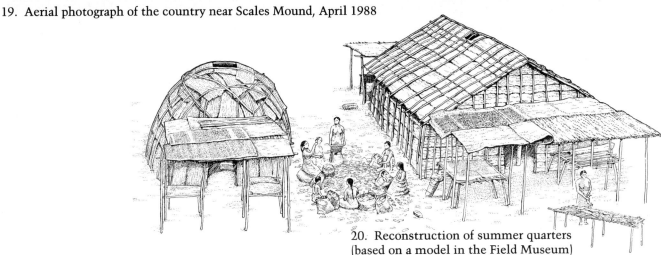

20. Reconstruction of summer quarters
(based on a model in the Field Museum)

The Later Indian Societies

WHEN THE FRENCH REACHED ILLINOIS in the late seventeenth century, they found the land peopled by an Algonquian-speaking confederacy of tribes, without any apparent links with the earlier peoples. The members of this Illinois Confederacy, which was dominated by the Illini, were not as sedentary as the Cahokians but took advantage of different habitats and seasonal opportunities for food gathering. Their effigy-mounds were much smaller than the mounds of the Cahokians.

The country around Scales Mound (figure 19), in the nonglaciated northwest corner of the state, gives us an idea of the variety of habitats that the Indians could choose within what is now Illinois; in the areas of wood and of prairie different forms of plant and animal life could be harvested at different seasons. The photograph reminds us that the term *mound* is ambiguous, for it can refer either to a constructed feature like Monks Mound or, as here, to a natural hill. Incidentally, it was often on such natu-

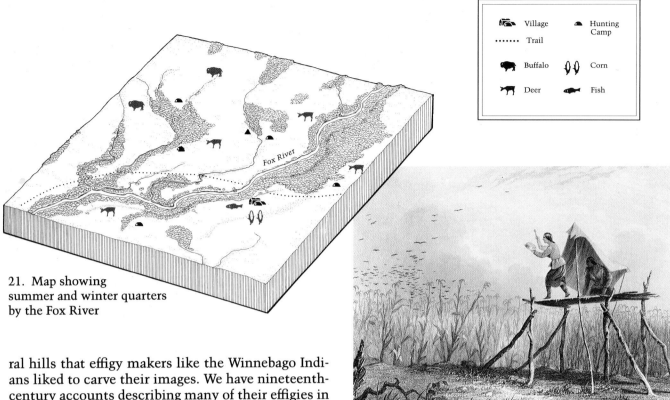

21. Map showing
summer and winter quarters
by the Fox River

ral hills that effigy makers like the Winnebago Indians liked to carve their images. We have nineteenth-century accounts describing many of their effigies in Illinois, but none of these sites now seems to be recognizable.

Most of the tribes in Illinois took full advantage of this variety of habitats. They liked to live during the summer in semipermanent villages, if possible by a stream or river, both for water and for fishing: there they could plant their corn fields in spring and summer. Figure 20 shows a typical Indian village at harvest time. The women are shucking the corn and gathering it into wicker baskets so that it can be dried and stored for the winter. This corn was crucial as a year-round food source, and the women's horticultural produce—beans, squash, and maize—was at least as important as what the men produced in their role as nomadic hunters.

Behind the women are the village houses, light structures with roofs and walls of bark, which could quickly be built from local materials. They could also be quickly abandoned, either when an enemy threatened or when it was necessary to move near new food sources. In winter small groups often set up camps in the shelter of the woods; from there the men went on hunting expeditions until it was time to tap the maple trees in the spring. In the early summer it was back to plant the corn, and the cycle began again.

The winter huts were low and covered with thick woven mats, to keep out the wind and offer some insulation from the cold; they were often dug some way into the ground. Winter was a time for hunting

22. Indians guarding the corn crop (Henry Schoolcraft, *The Indian Tribes of the United States*, Philadelphia, 1851–57; Newberry Library)

23. A herd of buffalo on the prairie (Henry Schoolcraft, *The Indian Tribes of the United States*, Philadelphia, 1851–57; Newberry Library)

and for elaborating those long stories in which verbal cultures are so rich. Figure 21 gives a summary indication of some of the sources of food for the later Indians on the Fox River. On the plains they hunted buffalo until most were killed by overhunting and hard winters at the end of the eighteenth century. In the wooded areas deer were abundant, and in the fertile river bottoms maize and wild rice were grown. The fish in the Fox River reminds us that, besides being farmers, the Indians were also skillful at fishing.

Indian Communications

WHEN THE STREETS OF CHICAGO ARE OUTLINED at night by the lights, someone in an airplane or a tall building sees a few marked diagonals, an obvious contrast to the main pattern of a north-south grid. Figure 24, taken from the top of the Hancock Building looking west, shows two of these pronounced diagonals near the horizon on the right. They are Elston Avenue and Lincoln Avenue and, like many of the other diagonal streets, are old Indian trails that have survived in the modern street pattern.

Figure 25 shows an aerial view of the Chicago region, drawn from information collected by the local archeologist Albert Scharf. According to Scharf's researches, a network of trails converged on the Chicago River mouth, and these can be identified with their modern continuation. Beginning in the northeast, we have the Green Bay Trail (Green Bay Road),

Little Fort Trail (Lincoln Avenue), Woodstock Trail (Elston Avenue), Rockford Trail (Lake Street), and so forth.

From the earliest times the Chicago portage, scene of the ancient runoff of glacial Lake Michigan, was an important link between the lake and the Des Plaines—Illinois river systems. This system came close to the coast of Lake Michigan at other points but only this one had a sheltered harbor, provided by the mouth of the Chicago River.

Many of the trails converging on the river went far into the country, and some were even statewide. For instance, the Green Bay Trail went right up to the town of that name. The Sauk and Fox Trail, which in places was worn a foot below ground level, ran almost straight out to Rock Island. Due south ran the Vincennes trail, first to Danville and then to Vincennes on the Wabash River, an important ford

24. Aerial photograph of the Chicago street pattern at night, 1989

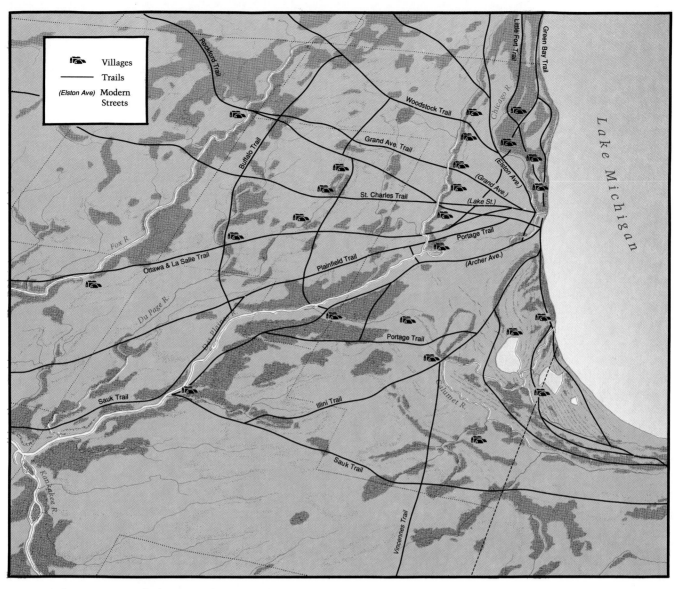

Villages

Trails

(Elston Ave) Modern Streets

Rockford Trail

Woodstock Trail

Grand Ave. Trail

St. Charles Trail

Buffalo Trail

Ottawa & La Salle Trail

Fox R.

Du Page R.

Des Plaines R.

Plainfield Trail

Portage Trail

(Grand Ave.)

(Lake St.)

(Elston Ave.)

Chicago R.

Little Fort Trail

Green Bay Trail

Lake Michigan

Portage Trail

(Archer Ave.)

Sauk Trail

Illini Trail

Sauk Trail

Vincennes Trail

Calumet R.

Kankakee R.

25. Aerial perspective of Chicago Indian trails (based on information from map by Albert Scharf, 1901, Chicago Historical Society)

and early French settlement (see chapter 3).

In figure 25 we have plotted not only the trails identified by Scharf, but also his Indian camp and village sites. If we consider those along the Des Plaines River, another important aspect of Indian communications becomes clear. Most of the trails run east-west; rivers took the place of north-south land trails, for the Indians were as much at home on water as on land. Any stream with water a few inches deep could carry their swift canoes. Thus the villages on the Des Plaines River had a much better north-south system of communications than a glance at the map might suggest. If we bear in mind the omnipresence of creeks and tributaries in Illinois, we can see that for the Indians the whole state was densely covered by an intricate system of communications.

26. Party of Winnebago descending the upper Mississippi River (George Catlin, *Souvenirs of North American Indians*, Newberry Library)

27. Striking the post, Indian preparations for battle (Henry Schoolcraft, *The Indian Tribes of the United States*, Philadelphia, 1851–57; Newberry Library)

Tribal Rivalries and Warfare

STARVED ROCK, on the Illinois River some thirty-five miles southwest of Chicago, is a site of extraordinary importance in the early history of the state. The great rock is one among a series of bluffs overlooking the river, carved by massive runoff from the forerunner of Lake Michigan. As figure 29 shows, it forms a natural fortress that is even more impressive when seen from the river below. When the French came to this country, they recognized the rock's importance and built Fort Saint Louis on it in 1682. This was for about two decades the stronghold not only for the French but also for the local Indians, who were by then menaced from the east by the Iroquois.

Long before the French arrived, the influence of the Europeans on the East Coast was disrupting Indian life in Illinois. Everywhere tribes were being seduced by the fur trade, which not only changed their subsistence patterns but also magnified traditional tribal rivalries into major conflicts. By 1650 the powerful Iroquois tribes in the East had acquired firearms from their British and Dutch allies and set about making war against the tribes further west, in order to control the lucrative fur trade.

The Illini were allied with the French, which made them a natural target for the Iroquois. During the second half of the seventeenth century the Iroquois began raiding in Illini territory, and Starved Rock became the great Illini bastion. By the first half

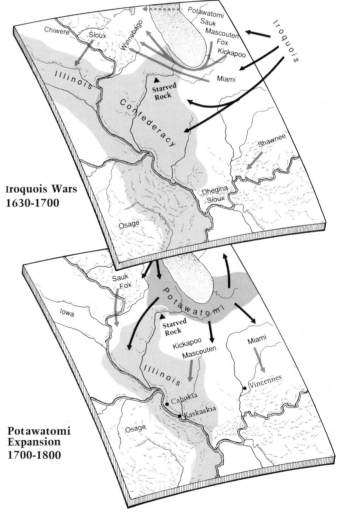

28. Maps showing tribal movements in the seventeenth and eighteenth centuries

of the eighteenth century the Iroquois threat began to subside, but other tribes that had earlier fled them began to press down into Illini territory. Chief among these were the Potawatomi, who occupied the whole southern end of Lake Michigan; they were joined by the Sauk and Fox, by the Kickapoos, and by the Mascoutens (figure 28). This confederacy tended to side with the British, while the Illini remained faithful to the French, and so the Indians of the region were caught up in the great colonial rivalry of the European powers.

Precisely what happened at Starved Rock and when are open to debate. But if we follow the tale told by N. Matson, an early Illinois settler who spoke with many Indians in the 1840s, it seems that in the fall of 1769 the combined Indian tribes attacked the Illini village near Starved Rock. After a series of bloody battles, about twelve hundred of the original three thousand Illini took refuge on the top of the rock. This group included about three hundred warriors, who for nearly two weeks resisted the repeated assaults of a force that was greatly superior in numbers. In the end hunger and tiredness won out, and all except one of the Illini perished, many by suicide. The accuracy of this appalling tale has been questioned by historians, but it seems certain that a band of Illini perished roughly in this way at Starved Rock towards the middle of the eighteenth century.

This was not quite the end of the Illini people, but they were certainly broken by the catastrophe. Many fled south to take up a miserable refuge near the French towns of the Illinois country, and by 1800 they had all but disappeared. The terrible massacre is hard to imagine today when you visit the delightful site of Starved Rock State Park.

29. Aerial photograph of Starved Rock, April 1987

The Arrival of the Europeans

B Y THE SECOND AND THIRD DECADES of the nineteenth century the position of the Indians in Illinois was becoming desperate. Until the time of the Revolutionary War, the British government, fearing military entanglements, had tried to limit European settlement to the eastern side of the Appalachian Mountains. After the Revolution, however, the settlers began to move westwards, at first in a trickle and then in a flood that plainly was going to overwhelm the Indians' hunting grounds east of the Mississippi River.

Indian leaders reacted in different ways to this threat. Black Hawk, a Sauk leader, eventually took up arms. In the Chicago region three notable Potawatomi leaders counselled accommodation with the

invaders. These chiefs were Shabbona, Sauganash (Billy Caldwell), and Chechepinqua (Alexander Robinson). The latter two were partly of European descent; Shabbona was an Ottawa Indian who married into the Potawatomi tribe. He fought with Tecumseh and the British against the republic in the War of 1812 but afterwards became reconciled to the advance of the United States. Thus he influenced the Potawatomi against joining the Winnebago uprising in 1827 and in 1832 opposed Black Hawk's offensive.

In the Treaty of Prairie du Chien in 1829, Shabbona, like Sauganash and Chechepinqua, received an allotment of land. His was about seventy miles due west of Chicago, at the spot now known as Shabbona Grove. This area was known for its splen-

30. Aerial photograph of Shabbona Grove, October 1987

31. Europeans view Cahokia Mounds (John Caspar Wild, *The Valley of the Mississippi*, St. Louis, 1841–42; Chicago Historical Society)

did trees and, as figure 30 shows, remains a most attractive site, now a state park. But Shabbona soon lost it to settlers through legal chicanery.

Indeed, virtually no members of the old society succeeded in coming to terms with the new one, so alien to them. Figure 31 shows two European gentlemen looking at the Cahokia Mounds, which settlers found so awe-inspiring. Figure 32 shows another extraordinary Indian work, the painting of the Piasa creature, which used to overlook the Mississippi River. Admired by Joliet, it was wantonly destroyed when the cliff was mined for lime—symbolic of cultural losses that can never be recovered.

The tribes do live on in the place-names of Illinois. We are accustomed to the idea that our major river names come from the Indian dialects—Mississippi, Ohio, Wabash, for example—and that the names of the great tribes have survived: Fox, Illinois, Potawatomi, Sauk, and so on. But it comes as a surprise to realize that the names of quite minor tribes are also commemorated in the place-names of Illinois: Cahokia, Kaskaskia, Peoria, Pickaway, Maroa, and Muscooten. So too are the names of many chiefs, in the locations named by the Europeans who took the land: Aptakasic, Ashkum, (Black) Partridge, Caldwell, Chebanse, Che-Che-Pinque, Du Quoin, Patoka, Pesotum, Sauganash, Senachwine, Shabbona, and Wilmette.

Then too some names refer (with varying degrees of confidence in the European translation) to local features: Channahon, Chicago, Hononegah, Nippersink, Pecatonica, Pecumsaugan, Piscasaw, Pistakee, Sinsinawa, Skokie, Somonauk, Watseka, and Waubonsee. All these names, which are only a sample of the full list, give us an idea of the extent to which Indian terminology has been perpetuated in our place-names.

32. Engraving showing the Piasa creature (Henry Lewis, *Das illustrierte Mississippithal*, Düsseldorf, 1854; Newberry Library)

3

THE FRENCH

33. Charles S. Winslow, Père Marquette's quarters by the Chicago River (Chicago Historical Society)

Introduction: French Sites in Illinois

THE FIRST EUROPEAN PENETRATION of North America was in areas very remote from Illinois—the Spaniards in the southwest, the French in the Saint Lawrence valley, and the English and Dutch on the eastern seaboard. Because of the interlocking water access provided by the Great Lakes from the Saint Lawrence valley, it was the French, or rather, Franco-Indian trappers, who first arrived in the Illinois country in the 1660s.

The first recorded passage through the region was by Louis Joliet and Père Marquette in 1673, during the voyage that gave rise to the map in figure 38. A phase of trading contacts followed, during which La Salle established forts at Peoria (1680) and Starved Rock (1682). Starved Rock had to be abandoned, but the Peoria fort survived (figure 34) and was succeeded in the early 1690s by a village and mission, the first European settlement in the state. Attempts to establish a post at the Chicago River had to be abandoned in the face of Indian hostility.

The next phase of French occupation involved not merely trading, but settlement and working of the land, and was focused in the southwest, on the east bank of the Mississippi River (figure 35). In 1699 a mission was established at Cahokia, close to the center of the ancient Mississippian culture; a village slowly grew around this mission, and four years later the same process began at Kaskaskia, which had the added advantage of nearby salt deposits. Slowly other centers emerged along the east bank on this flat and immensely fertile floodplain, which became the great food center for the French possessions in the whole Mississippi valley.

The growth of the English colonies on the eastern seaboard clearly posed a menace to this internal French development, and the French tried in the middle of the eighteenth century to contain English expansion by constructing a chain of forts (figure 56). However, the defeat of Montcalm at Quebec in 1759 meant that the Illinois country passed to the British by the treaty of 1763 and to the United States in 1783. The region for many years retained its French language and customs but was steadily absorbed into the English-speaking society that soon engulfed it.

34. Reconstruction of Fort Crevecoeur, Peoria (Illinois State Historical Library)

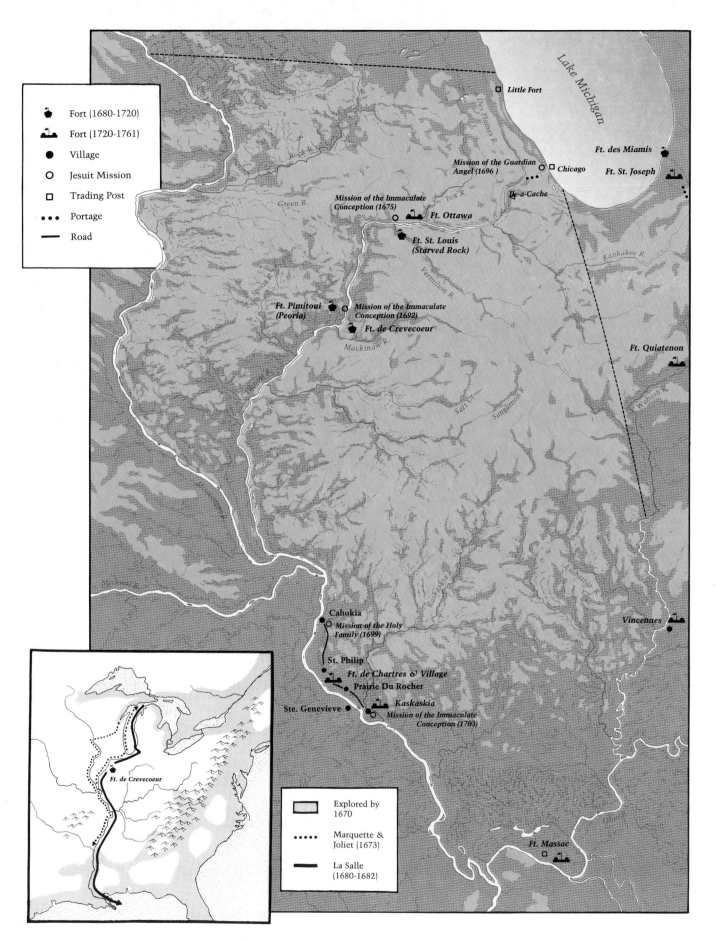

Legend (upper left):

- Fort (1680–1720)
- Fort (1720–1761)
- Village
- Jesuit Mission
- Trading Post
- ••• Portage
- —— Road

Lake Michigan

☐ Little Fort

Des Plaines R.

Rock R.

Ft. des Miamis

Mission of the Guardian
Angel (1696) ○ ☐ Chicago *Ft. St. Joseph*

Green R. Ile-a-Cache

Fox R. Kankakee R.

Mission of the Immaculate
Conception (1675) ○ *Ft. Ottawa*

*Ft. St. Louis
(Starved Rock)*

Vermilion R.

Spoon R.

*Ft. Pimitoui
(Peoria)* ○ Mission of the Immaculate
Conception (1692) *Ft. Quiatenon*

Ft. de Crevecoeur

Mackinaw R.

Salt Cr. Sangamon R.

Wabash R.

Illinois R.

Missouri R.

Kaskaskia R.

Embarras R.

Cahokia ●
Mission of the Holy
Family (1699) *Vincennes* ●

St. Philip ●
Ft. de Chartres & Village

Prairie Du Rocher

Ste. Genevieve ● *Kaskaskia*
Mission of the Immaculate
Conception (1703)

Ohio R.

Ft. Massac ☐

Inset map (lower left):

Ft. de Crevecoeur

Legend (inset):

- ▭ Explored by 1670
- ••••• Marquette & Joliet (1673)
- —— La Salle (1680–1682)

35. Map showing location of French sites in Illinois and
French exploration in North America, 1672–1760

36. Aerial photograph of Ile-à-Cache, October 1988

The Missionaries and Voyageurs

IT IS HARD TO FIND surviving material evidence from the earliest phase of French penetration. But this map (figure 38) seems to be one of those drawn for Governor Frontenac by Louis Joliet in 1674, upon his return from the Illinois country. Much of the map seems relatively crude, but the area around Chicago has been observed with remarkable detail and fidelity, perhaps because Joliet foresaw—with astonishing prescience—that this site was the key to the development of the whole region.

Leaving the "Lac des Illinois ou Missihiganin," the latter an early attempt at transliterating "Michigan," the traveller would enter the *havre*, or harbor, at what is now Chicago. The North and South branches of the river are shown, and off the South Branch comes the portage leading to the Illinois River, here described as the "Riviere de la Divine, ou

L'Outrelaize." The whole area was to be called Frontenacie, to please the powerful governor.

Note that Joliet includes information about local products. It may not be obvious how he learned that

37. Map of the Chicago portage in the early eighteenth century

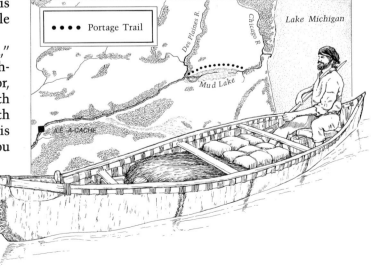

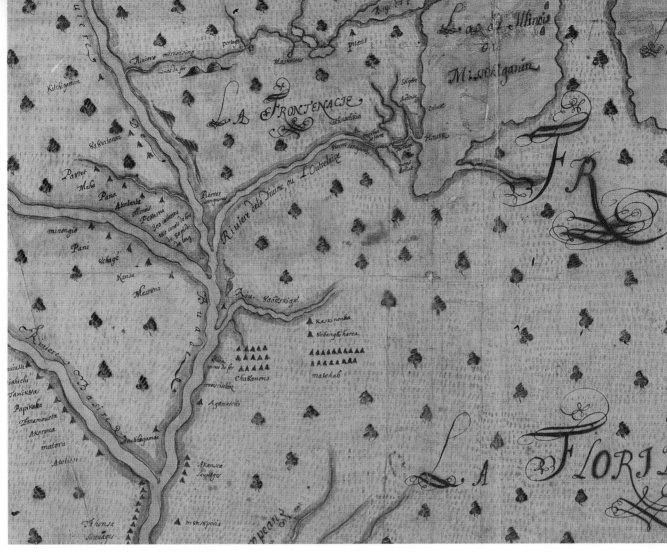

38. Detail from the Joliet map of 1674 (John Carter Brown Library)

copper (*cuivre*), slate (*ardoise*), and saltpeter (*salpestre*) were to be found just north of Chicago, but it was from personal observation that he located coal (*charbon de terre*) along the Illinois River just about where the massive pits would eventually be mined in the nineteenth century. From personal observation also he described the bluffs along the river as *pierres sanguines*, or reddish rocks.

A map like this may seem crude, but it was all that was necessary for a first penetration of the area between the Saint Lawrence and Mississippi valleys. The relations are roughly correct, as are the observations of major features; with a mental image like this, Canadian hunters and priests, and then traders, would over the next forty years circulate freely in this vast region. Most of what they saw in Illinois has passed away, but in the Des Plaines River by Romeoville is a little island, the Ile-à-Cache, which offers a faint echo of the time when it could have been a place for voyageurs to lie up and rest, somewhat safe from potentially hostile local Indians (figures 36, 37). The voyageurs would have come into the mouth of the Chicago River from Lake Michigan,

39. Early image of a beaver (La Hontan, *Mémoires de l'Amérique septentrionale*, Amsterdam, 1705; Newberry Library)

paddled down the South Branch, portaged across the mud lake, and paddled for some way down the Des Plaines River before reaching this sanctuary, where a small museum commemorates their former presence. (It should be added that documents establishing their presence at this precise site are not numerous.)

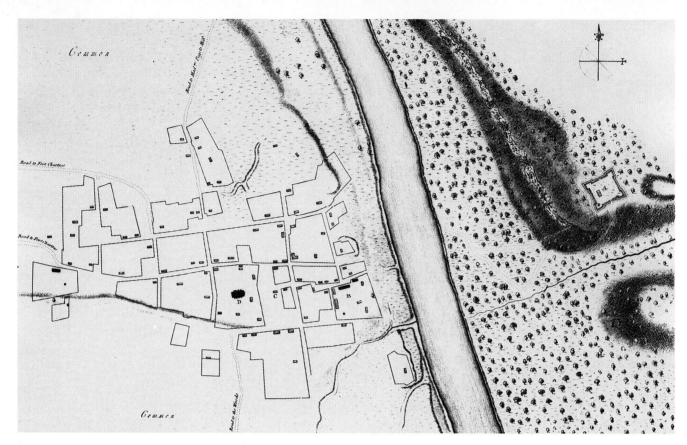

40. Captain Philip Pittman's plan of Kaskaskia, drawn about 1770 (*The Present State of the European Settlements on the Mississippi*, London, 1770; Newberry Library)

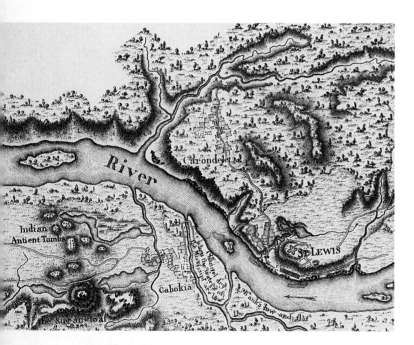

41. Plan of Cahokia (northward direction is to the right; detail from the "Map of the Country of the Illinois," in Victor Collot, *Voyage dans l'Amérique du Nord*, Paris, 1826; Newberry Library)

The Illinois Country

THE FOCUS OF THE FRENCH PRESENCE in Illinois soon moved from the north to the agricultural settlements along the east bank of the Mississippi River. Figures 40 and 41 are plans of the little towns at Kaskaskia and Cahokia, the latter hard by the "Indian Antient Tombs." Slowly other centers emerged along this strip of floodplain, and in 1717 the growing importance of the area was marked when jurisdiction over it passed from Quebec to the newly founded French settlement at the mouth of the Mississippi River, whose capital would become New Orleans. This southern settlement was of great strategic importance to the French, but it was chronically short of food, and the Kaskaskia-Cahokia region became the great breadbasket for the French in the Mississippi valley. Steadily the population grew, exporting to this assured market, until by 1750 there were about fifteen hundred inhabitants in the region, of whom perhaps five hundred were slaves.

Figure 43 is a satellite image of the Illinois country, covering roughly the same area as figure 42, a map by the English soldier-cartographer Thomas

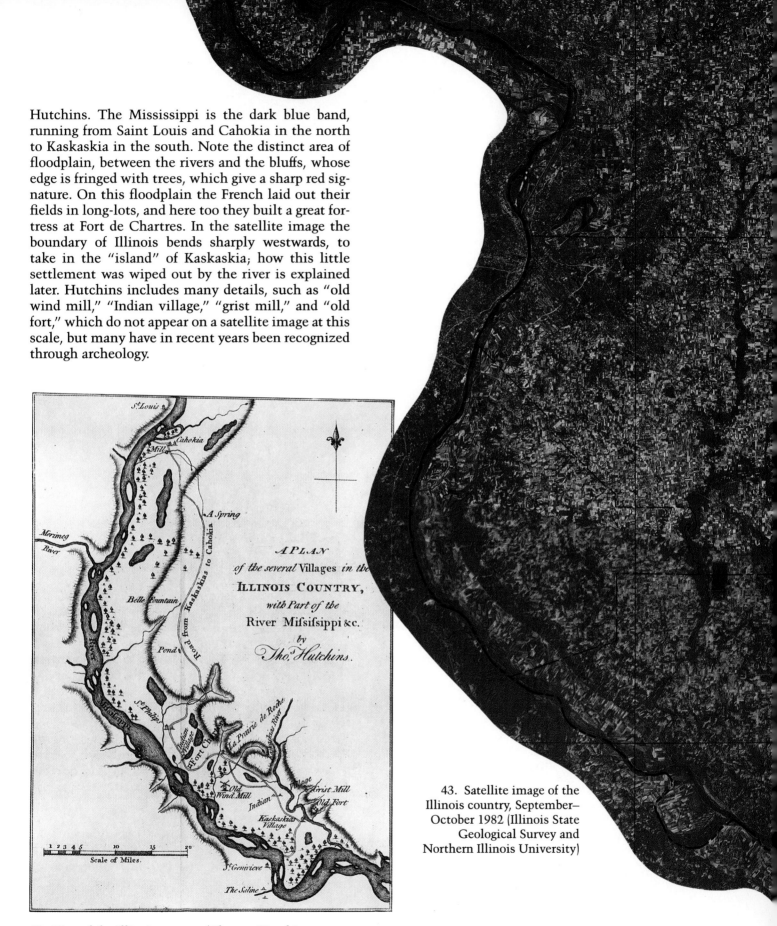

Hutchins. The Mississippi is the dark blue band, running from Saint Louis and Cahokia in the north to Kaskaskia in the south. Note the distinct area of floodplain, between the rivers and the bluffs, whose edge is fringed with trees, which give a sharp red signature. On this floodplain the French laid out their fields in long-lots, and here too they built a great fortress at Fort de Chartres. In the satellite image the boundary of Illinois bends sharply westwards, to take in the "island" of Kaskaskia; how this little settlement was wiped out by the river is explained later. Hutchins includes many details, such as "old wind mill," "Indian village," "grist mill," and "old fort," which do not appear on a satellite image at this scale, but many have in recent years been recognized through archeology.

42. Map of the Illinois country (Thomas Hutchins, 1778; Newberry Library)

43. Satellite image of the Illinois country, September–October 1982 (Illinois State Geological Survey and Northern Illinois University)

44. Aerial photograph of long-lots by the Wabash River, September 1986

Patterns of Landholding

EACH OF THE EUROPEAN PEOPLES that came to the New World had its own pattern of land settlement. The Spaniards, for instance, tended to allocate their land in large irregular rectangular units, suitable for ranches. The English, on the other hand, often used smaller units defined by the so-called metes-and-bounds system, which involved running boundaries from one convenient mark, perhaps a prominent tree or rock, to another; this gave very irregular shapes. The French in the Mississippi and Saint Lawrence valleys invented a distinctive system that makes use of the "long-lot."

The French made each grant at a right angle to some watercourse, with a relatively short width (normally one or two arpents) and a great length stretching into the backlands. Figure 45 shows a plan of some of these long-lots from the part of Illinois bordering on the Wabash River, near Vincennes in Indiana. This little town was established on the site of an Indian village by French traders in the early eighteenth century, and the best fields were across the Wabash River, in what would become Illinois.

The map was made in the 1820s as part of the township-and-range survey (on which see chapter 4); notice how the long-lots, with their Franco-Indian owners' names on them, contrast sharply with the squares into which the rest of the country was divided. These lots are somewhat reminiscent of the strips into which many arable fields were divided in medieval Europe, but the strips lacked the watercourse or transportation access that is an essential part of the French system in the New World. Figure 44 is a photograph looking west from a point above the Wabash River; three of the long-lots survive as spectacular intrusions into the normally rectangular pattern.

The greatest concentration of long-lots in Illinois was along the east bank of the Mississippi River in the area pictured in figure 43. They occupy virtually all the good land below the bluff, leaving the type of mark shown in figure 58. In fact, the long-lots can be discerned by careful examination of figure 43; it is rather extraordinary to think that the mark left by peasants on the Illinois countryside during the life of Louis XIV is still plainly visible from a satellite.

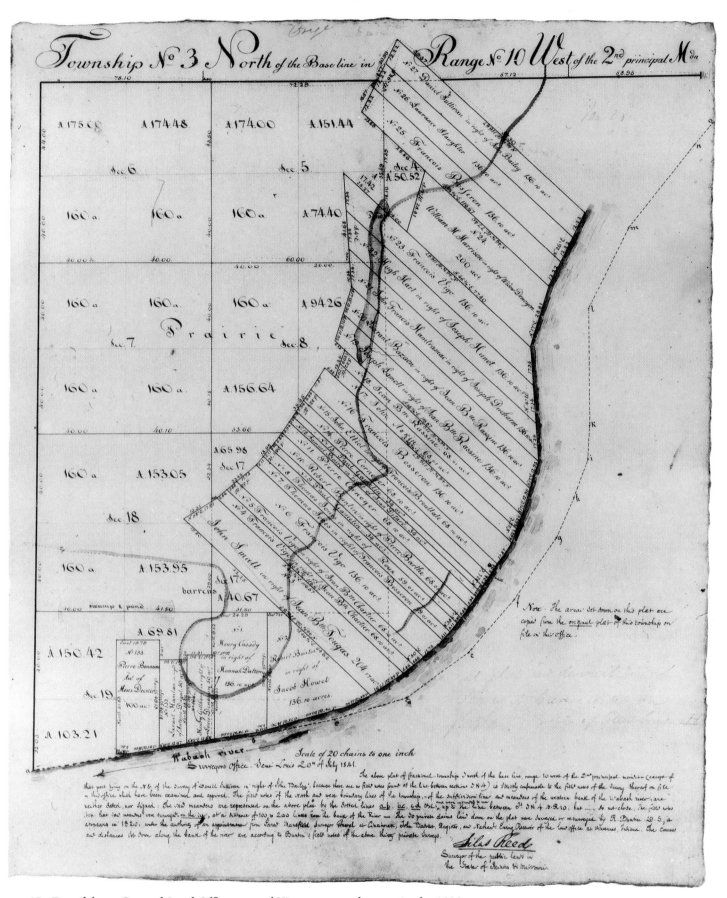

45. Detail from General Land Office map of Vincennes settlement in the 1820s
(Illinois State Archives, Springfield)

46. Aerial photograph of Cahokia church, September 1986

Village Life at Cahokia

CAHOKIA TODAY HAS ONLY TWO STRUCTURES dating from the eighteenth century: the church and the courthouse. Figure 46 shows the church, built in the mission grounds in the 1770s. A large modern church has been built immediately to the east of it, but the original structure, with its exposed timbering, gives us some idea of what public buildings looked like in the Illinois country. The courthouse has been much restored and often moved about, but seems to have begun life as a private house in the first half of the eighteenth century.

Figure 50 shows what the village of Cahokia might have looked like in 1760, based on Thomas Hutchins's plan of 1766 and on modern archeology. The houses straggled along the road for about three-quarters of a mile, with little thought of grouping for common defense, since the French were on good terms with the neighboring Indians. At the eastern edge of the village the fields stretched away, divided into long-lots which were dwarfed in area by the surrounding forest. A fence surrounded this *grand champ* to keep out animals, which were allowed to wander through the village. Other fences enclosed the house lots, and within these were elaborate gardens with vegetable plots, fruit trees, and even vineyards.

The houses themselves, as we find from surviving examples in Missouri (figure 48) and from contemporary illustrations (figure 49), were built using wooden uprights (sometimes sunk into the ground) to form a frame around a limestone chimney; an encircling covered verandah gave protection against the summer heat. Mud and clay formed the walls between the timbers, and these walls were normally plastered both inside and out, which gave them a tidy appearance. When the outside timbers were left unplastered, as on the church, these buildings looked startlingly like the half-timbered houses of Normandy, from which they may indeed derive.

All in all, life was relatively easy in the French

47. "Festivities of the early French in Illinois" (Henry Howe, *Historical Collections of the Great West*, Cincinnati, 1851; Newberry Library)

48. Photograph of Bolduc House, Sainte Genevieve, Missouri, May 1986

49. Early French home in Illinois valley
(Victor Collot, *Voyage dans l'Amérique du Nord*, Paris, 1826; Newberry Library)

Illinois villages. True, the Mississippi valley was known for its fevers and floods, but the land was as productive for the French as it had been for the mound builders. The country abounded in wild animals: buffalo, deer, and bear; the wolves too were so numerous that it was quite difficult to raise hogs outside pens. Wheat and maize were the main crops, and there were many lesser ones such as oats, barley, buckwheat, okra, tobacco, and onions. Some furs continued to be produced, but Illinois beaver skins were generally of poor quality. Salt came from the salt pans on the west side of the Mississippi, and there was an abundance of apples and other fruit. The market at New Orleans could never get enough of the Illinois foodstuffs; with the cash they got for their provisions the *habitants* could buy luxuries from France. They also bought slaves, whose number showed a slow but steady natural increase.

The influence of the French crown was remote, that of the church omnipresent. The parish priest was often the natural leader of the village, especially in the difficult days after 1763, and the whole community followed the rhythm of the church year, with its twenty-odd special holidays. The other important official was the lawyer, or *notaire*, for the members of the Illinois communities, like their counterparts in Normandy, seem to have been exceptionally litigious, always eager to establish and maintain their rights in court.

50. Aerial perspective of the village at Cahokia in the 1760s, from the northwest

1 Long Lots of the Common Field	5 Town Fortification
2 Barrier Fence of the Common Field	6 Cahokia Courthouse
	7 Blacksmith
3 Mission of the Holy Family	8 Road to Kaskaskia
4 Orchard of the Mission	9 Road to the Mississippi River

51. Aerial photograph of Fort de Chartres, September 1986

52. Aerial perspective of Fort de Chartres as it was in 1760

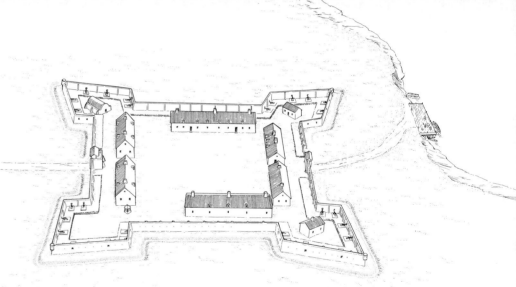

French Imperial Plans

IN SPITE OF THE USUAL FRIENDLINESS of the surrounding Indians, the French settlement in the Illinois country needed some kind of protection, and so in 1720 the first Fort de Chartres was built. It is described as "a fort of logs the size of a man's leg, square in shape, having two bastions which command all the curtains." For a long time its exact site was unknown, but in 1980 Terry Norris, district ar-

cheologist for the Saint Louis district of the U.S. Army Corps of Engineers, recognized its outline in an aerial photograph taken in 1928. Figure 53 shows this photograph, with the fort circled; it is a lozenge shape, with the bastions on the top right and lower left. Archeology has since confirmed this remarkable identification of the early fort.

It would have looked rather like Fort Crevecoeur,

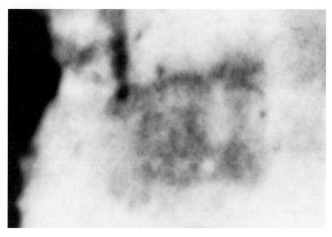

53. Aerial photograph with enlargement (right) showing earlier Fort de Chartres, 1928 (U.S. Army Corps of Engineers)

54. Infrared aerial image of Fort de Chartres, date unknown (Illinois Department of Transportation)

shown in figure 34, which was also made of logs. Such forts were very difficult to keep in good repair, and during the 1730s and 1740s there were frequent complaints about the condition of this fort and its wooden successor. By the early 1750s the Illinois settlement was of sufficient importance in the French imperial structure for a new and much larger fort to be envisaged. In 1753 the foundations were laid for a stone fort, which was to be part of the great chain of fortresses by which the French hoped to hem in the increasingly hostile anglophone colonies of the eastern seaboard (figure 56).

After many delays this fort was completed as shown in figure 52. It was roughly square, with living quarters on each side of the internal parade ground. At each corner was a bastion, a device where guns were placed to guard the adjacent wall; wherever an assailant came from, he would be caught in crossfire from at least two parts of the fort. In the top left-hand corner, within the bastion, was the powder magazine. Fort de Chartres was an odd structure, for it was not strong enough, with walls only two feet thick, to resist artillery bombardment (unless, as is possible, there once was an internal earthen platform running round it), yet it was excessively strong for resisting assault by Indians. Perhaps its value was largely symbolic, and it may have been designed in part as an economy measure; surely the stone walls would last longer than the wooden ones.

In the end this considerable work was rendered useless by the loss of Quebec in 1759; the French were obliged to cede Canada, and Fort de Chartres for a time became (British) Fort Cavendish. Eventually part was consumed by the river, and the rest fell into decay, from which it was rescued in this century by the Illinois Historic Preservation Agency. Figure 51 shows it in natural color on a fine summer day in 1986, and figure 54 in infrared false color during the winter. This latter figure, taken vertically, demonstrates the skill and accuracy with which the fort was laid out, though it is now lacking one side and two bastions.

55. Artist's impression of British soldiers entering
Fort de Chartres (Illinois State Historical Library)

56. Map showing New France and the war
with Britain, 1754–63

The Fall of New France

THE FRENCH POSSESSIONS IN THE NEW WORLD depended on France for their survival, and when the British seized control of the sea during the Seven Years War (1756–63), the French settlements' position became difficult. It became impossible after the fall of Quebec to the British in 1759, and by the Peace of Paris (1763) most of New France, including the Illinois country, passed into British hands. Figure 55 is an artist's impression of British soldiers beginning their occupation of Fort de Chartres. Many of the French *habitants* crossed to the west of the Mississippi, preferring to live in Saint Louis under Spanish rule.

In fact, the British influence on the Illinois country was minimal, for it was very remote from the center of their interests. The French in Illinois were much more seriously affected by the outcome of the War of Independence, for after 1783 they came under the rule of a nation bent on westward expansion and culturally alien to them. In almost every respect they were at odds with their new rulers. Mostly Catholic, they faced absorption by a mainly Protestant people. Law-abiding and even excessively litigious, they found it hard to deal with the lawless practices of the frontiersmen. Racially tolerant towards the indigenous people, they found it difficult

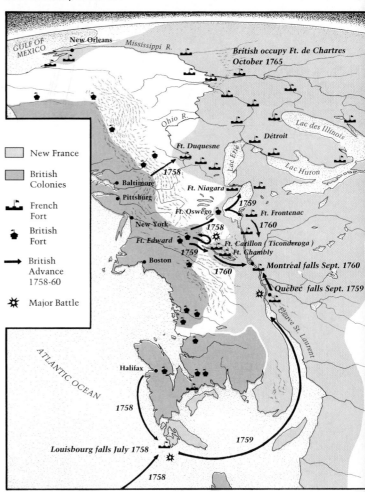

57. Aerial photograph of the Menard House, September 1986

to live alongside new immigrants who wanted to expel the Indians and take their land.

After a period of painful adjustment in the late eighteenth and early nineteenth centuries, the French were increasingly absorbed into the anglophone mass, profiting from the general peace and prosperity of the new republic, in whose affairs they eventually played an important part. One of the French leaders at this time was Pierre Menard, who became the first lieutenant governor of Illinois and whose charming house may still be seen down by the Mississippi, near the former site of Kaskaskia (figure 57).

This house epitomizes the style that the French developed in the Caribbean and carried into and up the Mississippi valley. It has a stone base, used for storerooms, and above it a main floor surrounded by a wide verandah. The numerous windows have louvers for ventilation, and two dormer windows are let into the low-pitched roof. At the back is a kitchen connected to the house by a covered way; this was a design intended to reduce the risk of fire. Houses in this efficient and appealing style may be found all over the area of French influence, from the hills of Saint-Domingue (now Haiti) to the lowlands of the Mississippi.

Up on the hills behind the house may still be seen the remains of the eighteenth-century fort, shown in the plan of Kaskaskia (figure 40); this is part of Fort Kaskaskia State Park. It is possible to trace in the grass the outlines of a bastionned square rather like the one in figure 54, but the many trees on the site make it almost impossible to photograph it from the air, even in winter.

58. Aerial photomosaic of East Saint Louis, 1958 (University Library, Champaign-Urbana)

The French Villages Today

ALTHOUGH ONLY TWO EARLY BUILDINGS SURVIVE at Cahokia, the area of the village is proving to be extremely abundant in archeological remains, some of which can be collected on the present surface, and it profoundly marked the existing layout of East Saint Louis. Figure 58 is a photomosaic of this town and the area to the south of it (photomosaics are made by overlapping a number of aerial photographs taken on successive runs). We have marked Cahokia, lying right by the river on the edge of a housing development.

Most striking is the way in which the alignment of the whole of this part of town has been influenced by the southeast-slanting great common field of the original village. As the *Standard Atlas of Saint Clair County* (Chicago, 1901) shows, the common fields of Cahokia and of Prairie du Pont, an adjoining vil-

lage, were very extensive, so that it is not surprising they ended up cutting a great gash into the conventional north-south checkerboard; this is visible even in the satellite image, figure 43.

Kaskaskia, on the other hand, left very little mark. When the British army captain Philip Pittman wrote about it in 1770, he described it as "by far the most considerable settlement in the country of the Illinois, as well from its number of inhabitants, as from its advantageous situation." It then had sixty-five European families, as well as some more or less transient merchants and a considerable number of slaves. As figure 40 shows, it was informally planned, on the west bank of the River Kaskaskia, but it had a stone church, a stone house and chapel for the Jesuits, and some other stone houses. Large boats could tie up at its wharf, safe from the

59. House at Kaskaskia falling into the river about 1900 (Illinois State Historical Library)

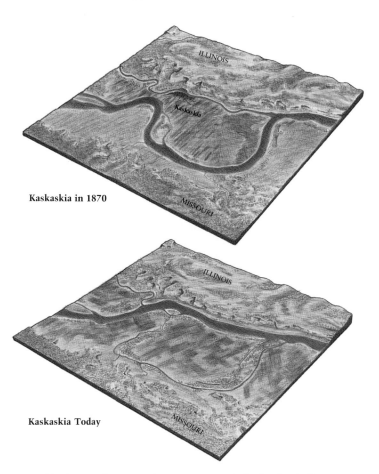

Kaskaskia in 1870

Kaskaskia Today

60. Maps showing the sequence of Kaskaskia's disappearance

tempestuous Mississippi River, and the Kaskaskia River also allowed the establishment of water mills.

However, this riverine site was Kaskaskia's undoing. As figure 60 shows, only a narrow neck of land separated the Kaskaskia River from the Mississippi, at a point where the great river makes an abrupt right turn. Over the years the Mississippi gnawed away at this neck, until during a great storm in 1881 it finally broke through, washing the neck of land away and moving into the channel of the Kaskaskia River. During the following years the village of Kaskaskia was utterly destroyed, as figure 59 illustrates; only the Menard House survived, up on its bluff. Today, the Mississippi flows close under the bluffs, and the Kaskaskia common fields, on which the long-lots are still visible (figure 43), lie across the river—though still in Illinois.

4

ANGLOPHONES IN
THE SOUTH

Introduction

DURING THE PERIOD OF BRITISH CONTROL, from 1763 to 1783, very few English speakers found their way from the East into the Illinois country. They all came obscurely, many on foot and often alone. After independence, however, whole groups of people began to cross the Alleghenies, often bringing their possessions with them in carts. After 1792 many of the one hundred thousand people of European stock in Kentucky entered southern Illinois near the confluence of the Ohio and Wabash rivers.

Federal statesmen like Thomas Jefferson were well aware of this movement and had done the best they could to ensure that it took place in an orderly fashion. In 1785 the Land Ordinance was passed, regulating the way in which the land, once cleared of Indian claims, would be divided up. Two years later came the Northwest Ordinance, which made provision for the disposal of the land on a larger scale. When an area had five thousand inhabitants, it could apply to Washington for recognition as a territory, and when the population reached sixty thousand, full statehood could be requested. Maps of the period show this process in full swing, as first territories and then states appeared further and further to the west: Kentucky in 1792, Ohio in 1803, Indiana in 1816, and Illinois in 1818 (figure 61).

The Illinois country was captured for the republic by George Rogers Clark in 1778, but it was not until 1787 that General Harmar, a federal representative, arrived at Kaskaskia. He found the area in considerable disorder, for the British were disinclined to abandon their interests and encouraged Indian resistance to the settlers from the East. By 1800 the region had only about twenty-five hundred inhabitants, the same number as in 1750. However, between 1795 and 1809 the Indian tribes, who after a number of battles (figure 62) were forced to vacate their lands, signed treaties that opened the way for an initial wave of settlement, which carried the area's population to 12,282 by 1810.

In 1811 the first steamboat made its way down the Ohio to the Mississippi River, and with the end of the War of 1812 the great rush was on. Settlers came from the South up the Mississippi, and from the East along the Ohio, pouring into southern Illinois through centers like Shawneetown, which for a while enjoyed great prosperity. Land offices in Kaskaskia, Shawneetown, and Vincennes were enormously active in selling plots in the new territory, which by 1818 had enough inhabitants to become the state of Illinois. Almost all these settlers came from southern states, though immigrants came as well from England, like the celebrated Birkbeck. As yet, virtually none came through northern Illinois.

The new arrivals usually sought timbered land within easy reach of a river: the fertility of the open prairies was not yet appreciated, and the riverboat was by far the best way to bring goods in and to take produce out. Settlements like New Salem appeared far up what look to our eyes like hopelessly narrow and shallow rivers, as the tide of pioneers began to fill up the southern part of the state. Land transport was for a time confined to the old Indian trails, but by the 1830s such routes as the National Road from the East were being opened (figure 62). The era of mass long-distance travel by coach was short, however, for in a decade or two the railroads would replace most coaches and steamboats.

61. Detail from *A New Map of Part of the United States,* by John Cary (London, 1825; Newberry Library)

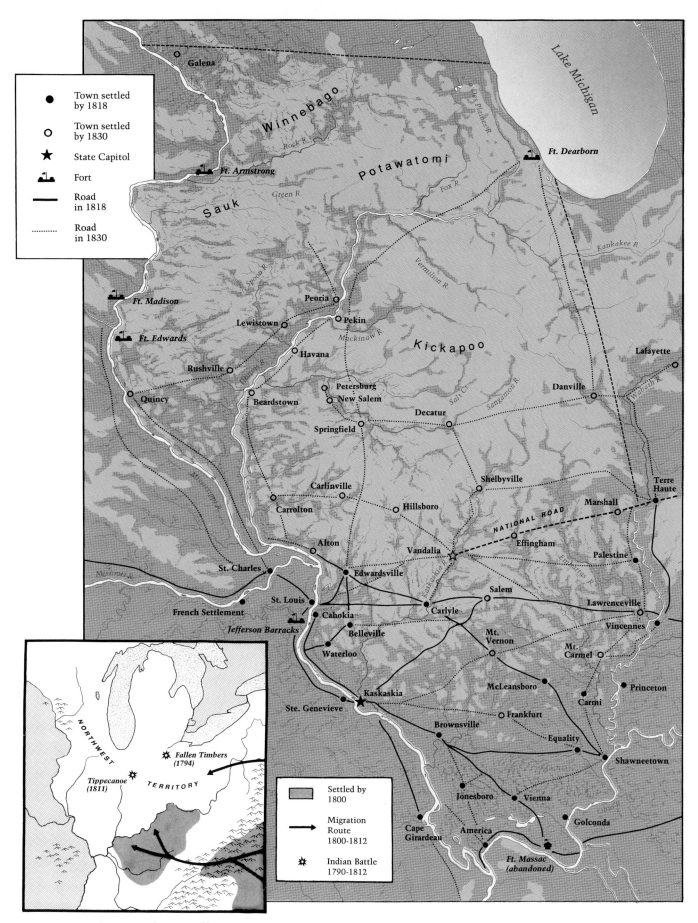

Legend

- ● Town settled by 1818
- ○ Town settled by 1830
- ★ State Capitol
- ⚓ Fort
- ▬ Road in 1818
- ⋯ Road in 1830

Lake Michigan

Galena

Winnebago

Rock R.

Ft. Armstrong

Sauk

Green R.

Potawatomi

Des Plaines R.

Ft. Dearborn

Fox R.

Kankakee R.

Ft. Madison

Spoon R.

Peoria

Lewistown

Pekin

Mackinaw R.

Kickapoo

Vermilion R.

Lafayette

Ft. Edwards

Havana

Rushville

Illinois R.

Quincy

Beardstown

Petersburg
New Salem

Springfield

Decatur

Salt Cr.

Sangamon R.

Danville

Wabash R.

Carlinville

Shelbyville

Terre Haute

Carrolton

Hillsboro

Marshall

NATIONAL ROAD

Alton

Vandalia

Effingham

Palestine

Embarras R.

St. Charles

Missouri R.

Edwardsville

Kaskaskia R.

Salem

Lawrenceville

St. Louis

Cahokia

Carlyle

Vincennes

French Settlement

Jefferson Barracks

Belleville

Mt.
Vernon

Mt.
Carmel

Waterloo

Kaskaskia

McLeansboro

Princeton

Carmi

Ste. Genevieve

Frankfurt

Brownsville

Equality

Shawneetown

Jonesboro

Vienna

Cape
Girardeau

America

Golconda

Ft. Massac
(abandoned)

Ohio R.

Inset map

NORTHWEST

Fallen Timbers
(1794)

Tippecanoe
(1811)

TERRITORY

- ▓ Settled by 1800
- → Migration Route 1800–1812
- ✱ Indian Battle 1790–1812

62. Map showing Illinois region, 1781–1830

Detail of a Township

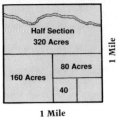

6	5	4	3	2	1
7	8	9	10	11	12
18	17	16	15	14	13
19	20	21	22	23	24
30	29	28	27	26	25
31	32	33	34	35	36

Detail of a Section

Half Section
320 Acres

160 Acres | 80 Acres
| 40

1 Mile

1 Mile

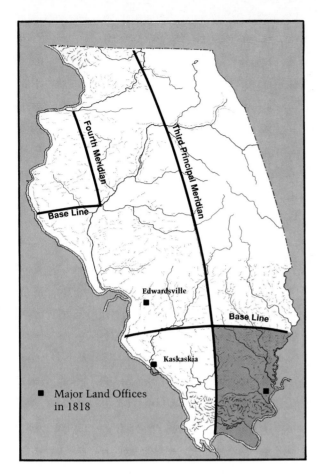

63. Diagram of the township-and-range system

Area of Townships

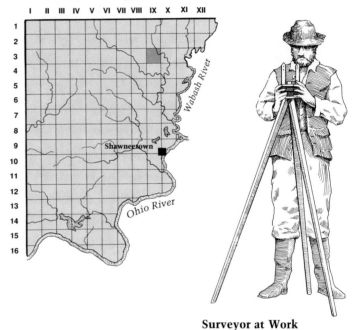

Surveyor at Work

The Township-and-Range System

THE BASIC TERRITORIAL UNIT chosen for the Land Ordinance of 1785 was the six-mile-square township, each divided into thirty-six one-mile-square sections. These townships were keyed into a general system of north-south meridians and east-west baselines. The survey began in Ohio in the 1780s and, working constantly westwards, by 1804 reached the boundary of the Illinois Territory, near Vincennes. There the surveyors had to contend with existing long-lots (shown in figure 44). but they followed their instructions and arranged their quadrilaterals around these intrusions.

As figure 63 shows, the Illinois Territory had a convenient meridian more or less down the center of it, and a baseline across the bottom quarter. As the first detail in this diagram makes clear, the townships were located by being, say, 1 south of the baseline and 2 east of the (third principal) meridian. Within the township the sections were numbered in sequence beginning at the top right and ending at the bottom right. Each section was a mile square, which gave it 640 acres; a half section made up a farm of very reasonable size, on good land.

Eventually the whole of the state was laid out in this way, giving it an astonishingly geometrical appearance from the air. Figure 64 shows the region alongside the Wabash and Ohio rivers, where the settlers were pushing into the state and where the early surveys were made. The system was relatively easy to lay out in this flat country, and today we are astonished by the accuracy with which these surveyors could lay a line over hundreds of miles, on foot, with no recourse either to aerial photography or to astronomical fixes. As the earth is not flat, from time to time the surveyors had to put in "offsets," which made up for the shrinking size of the squares as they moved northwards (see figure 283). The land thus staked out was sold at land offices

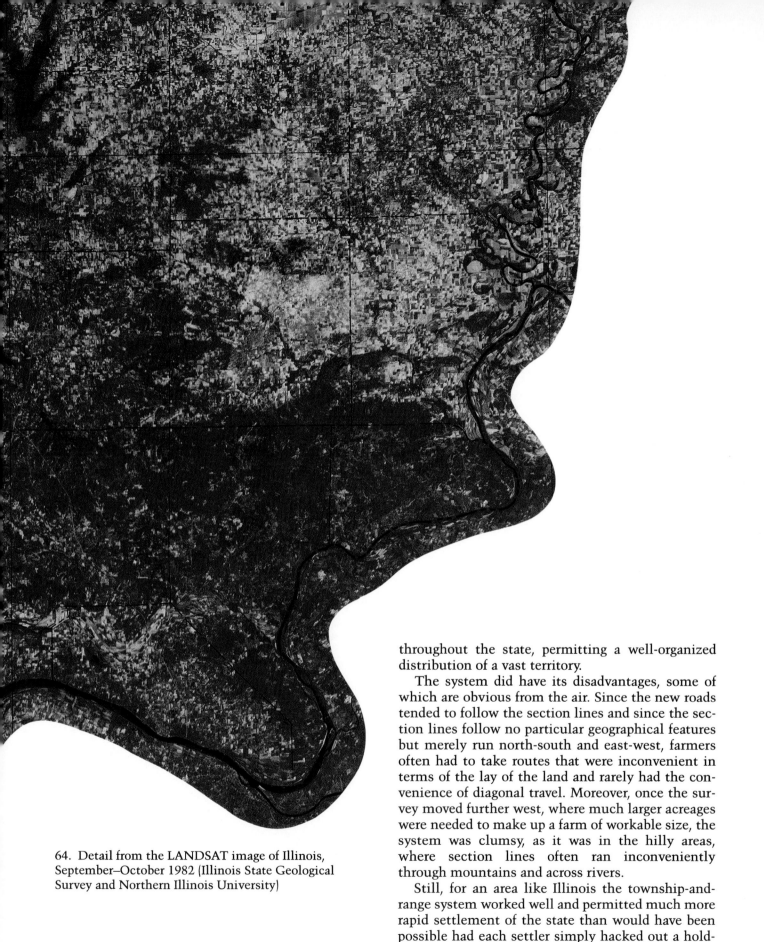

64. Detail from the LANDSAT image of Illinois, September–October 1982 (Illinois State Geological Survey and Northern Illinois University)

throughout the state, permitting a well-organized distribution of a vast territory.

The system did have its disadvantages, some of which are obvious from the air. Since the new roads tended to follow the section lines and since the section lines follow no particular geographical features but merely run north-south and east-west, farmers often had to take routes that were inconvenient in terms of the lay of the land and rarely had the convenience of diagonal travel. Moreover, once the survey moved further west, where much larger acreages were needed to make up a farm of workable size, the system was clumsy, as it was in the hilly areas, where section lines often ran inconveniently through mountains and across rivers.

Still, for an area like Illinois the township-and-range system worked well and permitted much more rapid settlement of the state than would have been possible had each settler simply hacked out a holding, as happened in parts of the East.

65. Aerial photograph of the bank at Shawneetown, September 1986

Shawneetown and the Ohio River

IN 1815 THE FOCUS OF NEW SETTLEMENT AND COM-
MERCE in the Illinois Territory was the Wabash
River valley. But this was a backwater compared
to the Ohio River, where thousands of farmers and
scores of small towns were establishing themselves
from Indiana to Pennsylvania. It was natural that the
early life of Illinois would be tied to this greater
river, freely accessible from the East. The federal
government in 1812 selected the site of Shawneetown, at the confluence of the two rivers, for its local
land office. The office opened there in 1814, and the
town was duly platted in accordance with instructions from Washington—a rare example of a feder-
ally planned town.

Shawneetown quickly grew as a center for land
sales and as a port for the growing numbers of flatboats carrying Ohio River agricultural produce to
New Orleans. By 1819 the thriving town had about
a hundred houses and some churches and banks.
The salt industry, which had existed in a small way
in French and Indian times, was producing three
hundred thousand bushels annually and exporting it
by river, south to New Orleans. On the river also
came a great tide of immigrants into southern Illinois, often in flatboats like the ones shown in figure
66. There were so many immigrants that they quite
overran the ferry, and in 1818 a traveller recounted
how he waited for the best part of the morning in
order to get across the river. These migrants, together with the salt and banking industries, seemed
to be setting Shawneetown on the way to becoming
a substantial city.

However, the site was from the start plagued by
floods during the annual spate of the Ohio River. As

is usual in this kind of floodplain, the land tended
to slope back from the levee, so that the waters crept
around from the back and cut the town off from its
hinterland for quite long periods. As early as 1817,
when Morris Birkbeck passed through Shawneetown, he remarked, "This place I account as a phenomenon, evincing the pertinacious adhesion of the
human animal to the spot where it has once fixed
itself," so inconvenient did he find the floods.

If Shawneetown did not grow as fast as had been
hoped, at least it survived, and in 1824 it was the
meeting point for at least four stagecoach mail
routes, running to Vincennes, Kaskaskia, Golconda,
and McLeansboro. Throughout the nineteenth century the Ohio River continued its seasonal ravages,
to which the inhabitants of Shawneetown became
more or less accustomed. In 1937 came huge floods
that swept right through the town; a new settlement
was established on the hills some way back from the
river, and many inhabitants left what was called Old
Shawneetown.

Today it is still possible to imagine the little town
at the height of its influence. The streets have retained the rectangular plan that the surveyors gave
them in the early nineteenth century, and the Bank
of Illinois Building, shown in figure 65, speaks
plainly of the early aspirations, even if the lots
around it are now largely vacant. We need not believe the story that Shawneetown bankers once re-
fused a loan to the infant village of Chicago on the
grounds that, not being on a sizeable river, it would
never amount to much. But it is a good story, and
one that captures the realities of power in the early
nineteenth century.

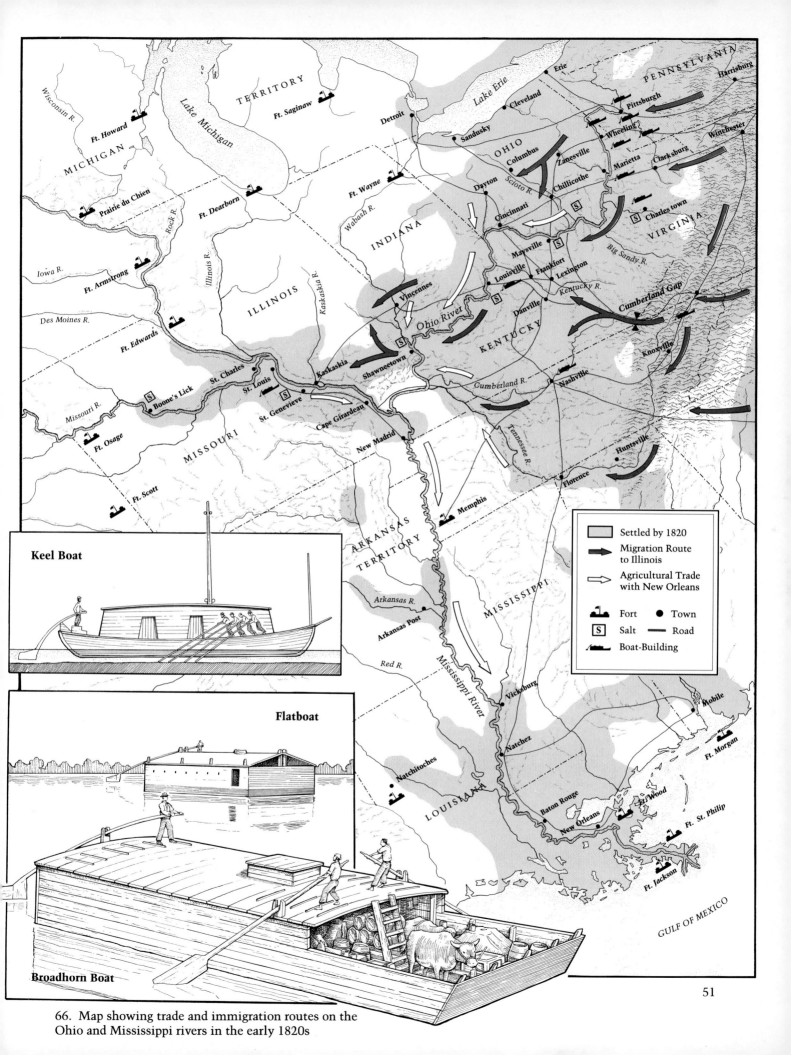

66. Map showing trade and immigration routes on the
Ohio and Mississippi rivers in the early 1820s

67. Aerial photograph of meanders on the Wabash River, September 1986

An Incongruous English Settlement on the Prairies

IT IS SOMEWHAT IRONIC, given British intransigence over federal claims in the Illinois country, that one of the earliest and most influential settlers came from England. Morris Birkbeck was the son of a rich Quaker, but England in 1815 seemed to have little to offer him. The economy was in a slump, with widespread unemployment following twenty years of war, and the political situation seemed bleak; the monarchy was widely disliked, and England was aligned on the Continent with the reactionary powers that had eventually succeeded in defeating republican France.

Birkbeck, like many others, dreamed of starting fresh. He associated himself with a younger man, George Flower, who in 1816 visited the United States. Flower came well recommended, with a letter of introduction from Lafayette, and after travelling as far as Tennessee and Illinois spent part of the winter with Thomas Jefferson at Monticello. In the spring of 1817 Birkbeck joined him, and they travelled westwards. Crossing Indiana, they took the ferry across the Wabash near Harmony. The river wound as sinuously as in figure 67, but the bottomlands were much more encumbered with vegetation, through which Birkbeck and Flower for some time struggled. Suddenly they emerged on their first prairie. Birkbeck was transported; as he later wrote, these open areas were "so beautiful with their surrounding woods, as to seem like the creation of fancy; gardens of delight in a dreary wilderness."

Birkbeck and Flower became excellent customers for the land office at Shawneetown, entrenching themselves at English Prairie, shown in figure 69. They had hoped to establish distinctive communities on the prairie, but the English farmers, artisans, and laborers they imported did not find things easy, in spite of the uncommon precautions that Birkbeck

68. George Flower, Park House, Albion, ca. 1820 (Chicago Historical Society)

69. Detail from *Map of Illinois*, by John Melish (Philadelphia, 1818; Newberry Library)

took in providing healthy housing and medical services. The prairie was very difficult to cultivate, the insects were troublesome, and water was scarce. Their neighbors, rude pioneers from the South, were not particularly friendly towards the English, who were seen as overfriendly towards their late allies, the local Indians, and as addicted to overgentlemanly pastimes such as playing cricket, riding to hounds, and holding agricultural fairs.

Eventually Birkbeck's settlement at Wanborough totally failed, but Flower's Albion knew a mild prosperity and survives to the present day. Birkbeck's greatest contribution to his adopted state were his publications extolling the potential of the prairies: *Notes on a Journey from the Coast of Virginia to the Territory of Illinois* (1817) and *Letters from Illinois* (1818) went through many editions in English and were translated into German. In spite of the counterpropaganda mounted by the celebrated English journalist William Cobbett, these works no doubt played their part in drawing settlers to the new state. Birkbeck was also active in local politics and in 1824 was prominent in defeating the move to amend the new state's constitution in order to permit slavery.

70. Early pioneers
in Illinois (Illinois State
Historical Library)

New Salem: Settling the Center of the State

ABOUT THE TIME that Birkbeck was trying to establish himself near the Wabash River, in the center of the state the northernmost tide of settlement began to reach up the Illinois River and eastwards along the valley of its tributary, the Sangamon River. These settlers, mostly from southern states, sought well-timbered valleys where they could find wood for their houses and fences, and water both for drinking and for power mills.

In 1829 two pioneers from Georgia, Rutledge and Camron, thought they had found an excellent mill site at a place on the Sangamon River that they called New Salem. There they built a gristmill, which was soon followed by a store, a saloon, and a post office, and before long a little town had come into being (figure 72). New Salem is known to us today mainly because Abraham Lincoln spent six formative years there, but the town itself gives us a fascinating picture of a settlement of the 1830s.

The prairie nearby was potentially productive, but the settlers at New Salem, most of whom were from the South, were no more skillful than Birkbeck's laborers at breaking it up for tillage. They had emerged from the woods of Kentucky and Tennessee with a certain mode of cultivation, and in Illinois they continued to think in terms of small mixed farms that could be largely self-sufficient. They grew most of their own food and for professionals like doctors and teachers often relied on Yankee immigrants, who began to enter the country with new skills learned in the states of the northeast.

71. Map of the neighborhood of New Salem, about 1835

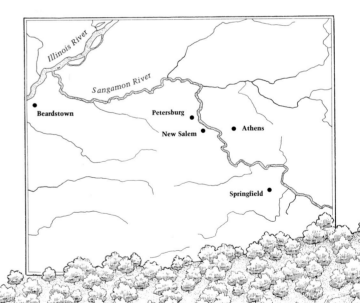

1 Onstot's Cooper Shop
2 Barn
3 Miller, Blacksmith
4 School

The site was soon beset with difficulties. Although in favorable weather a steamboat could get up the Sangamon River quite far past New Salem, during many months of the year the river was impassable for anything heavier than a canoe. New Salem did have communications with Springfield by stagecoach after 1834, for it was on the line to Havana, but as time went by Petersburg, better sited on the river, came to provide the services that the farmers of the area needed. From about 1836 New Salem began to decline, and by 1843 it had been abandoned.

The renaissance of New Salem began in 1906, when publisher William Randolph Hearst bought the area. It eventually passed to the state, which undertook the restoration of the site and the reconstruction of the buildings, using the 1829 plat that was found in the Sangamon County archives. Today visitors can inspect the mill on the river, the nearby store and saloon, and the faithfully reconstructed houses, getting a good idea of the birth and development of the town. The area is still quite heavily wooded, and the houses are widely dispersed, allowing for substantial fenced plots around them.

73. Ground view of New Salem, September 1988

72. Reconstructed aerial perspective of New Salem in the 1830s

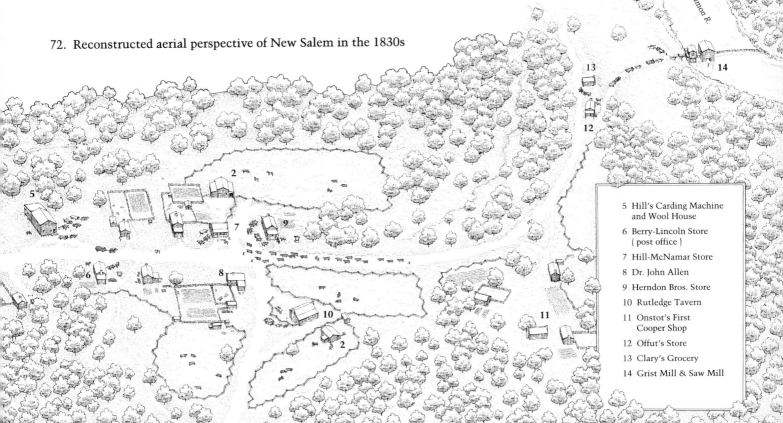

5 Hill's Carding Machine and Wool House

6 Berry-Lincoln Store (post office)

7 Hill-McNamar Store

8 Dr. John Allen

9 Herndon Bros. Store

10 Rutledge Tavern

11 Onstot's First Cooper Shop

12 Offut's Store

13 Clary's Grocery

14 Grist Mill & Saw Mill

74. Aerial photograph showing the line of the National Road, 1972 (University Library, Champaign-Urbana)

The Emerging Road Network

MANY IMMIGRANTS REACHED THE WEST on foot or on flatboats. A new and slightly more comfortable possibility opened up, once stagecoaches began to run on a number of routes. Figure 76 shows the road network at it was in the 1830s. The very large road coming in from the East was the National, or Cumberland, Road, which began in the East in 1811 and by 1839 reached Vandalia. It went no further under federal auspices, because the emergent state of Illinois questioned the right of Washington to determine the line of roads within the state. Nevertheless, in an era when the young state's river systems tied her to the South, it was an important link with the East, and a well-travelled immigrant route.

This National Road was succeeded by U.S. 40, which runs on much the same alignment. As our aerial photograph of the section near Greenup (Cumberland County) shows, the road ran impressively straight, in disregard of the quadrilateral township system. Indeed, it ran much straighter than does its successor the expressway just to the north, or the old Saint Louis, Vandalia, and Terre Haute Railroad to the south. Towns like Greenup were aligned to the old road, along which many sturdy stone bridges may still be found.

Many of the early roads were known as "traces." If they were anything like the early roads in Europe, they were very informal in their structure. Streams and rivers were crossed either by expensive bridges

75. Photograph of the Clayville Tavern, September 1988

or, more often, by fords, and crossing marshy patches the road often became very wide, as drivers sought firm ground on either side of the main track. In 1830 no road ran up the valley of the Illinois River to Chicago, no doubt because the ground was so marshy in that area and the little outpost was not yet important. Large areas in the center of the state were far from any coach route. This did not change much before the 1850s, when the all-conquering railroads penetrated every corner of the territory.

Coaches could go only a limited distance each day. This meant that passengers often had to get off the coach—which they must have been glad to do, given the state of the roads and the absence of effective springs—and find lodgings for the night. Throughout the state there are many houses of the second quarter of the nineteenth century that once served as wayside inns. Some are identified and some are not; figure 75 is a photograph of the charming Clayville Tavern just outside Springfield.

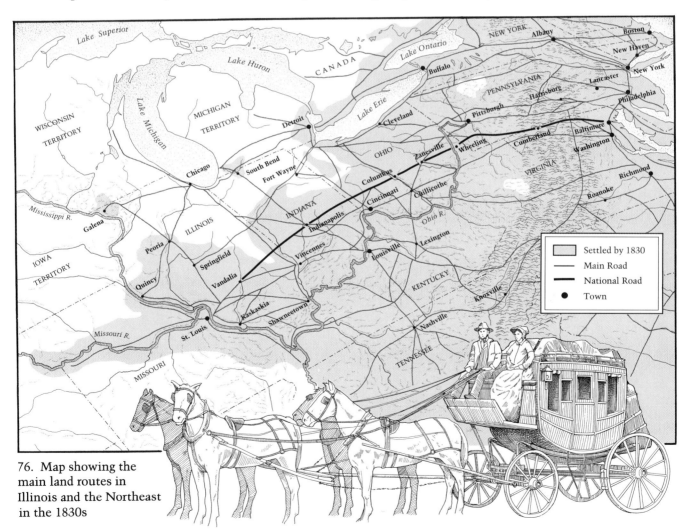

76. Map showing the main land routes in Illinois and the Northeast in the 1830s

77. Aerial photograph of a steamboat on the Mississippi River, September 1986

78. River view (Friedrich von Hellwald, *Amerika in Wort und Bild*, Leipzig, n.d.; Newberry Library)

Steamboats Revolutionize the Waterways

IN THE TIME OF THE INDIANS, every rivulet was a potential highway for their canoes. Flatboats could also venture far up narrow streams and carry substantial cargoes. Flatboats had great disadvantages, however, for they could travel easily only downstream, at the speed of the current. Moving upriver was a laborious process for teams of men poling the boats along the banks. Thus the produce of Illinois could easily pass down to New Orleans on the Mississippi, but returning goods upriver was often prohibitively expensive.

From about 1810 onwards, a new form of water transport spread rapidly on the rivers of Illinois: the steamboat. These rivercraft were a revolution on the waterways; they moved much faster and allowed setting up and maintaining precise schedules, which was critical for commerce. Steamboats were constructed either as side-wheelers or stern-wheelers, and each type had its advantages. Side-wheelers were more efficient and swift on the main watercourses, as suited a deeper draft and higher operating costs. Stern-wheelers came into their own on the narrow tributaries. They could carry much larger loads and go much faster than the flatboats, and their shallow draft allowed them to penetrate far into the backwaters. The *Talisman*, which Lincoln occasionally helped pilot, went up the Sangamon River as far as Petersburg, and other steamboats ventured up the Kaskaskia River as far as Vandalia (figure 79).

Side-wheelers mostly operated on the Mississippi River south of Alton, though special lighter-draft vessels of this kind were also developed for shallower waters; in figure 78 a side-wheeler is loading cargo right off the bank of the Ohio River. Saint Louis and to a lesser extent Alton became the great transfer ports where cargo from the bigger boats of the lower river was shifted onto the lighter-draft boats of the upper river. Saint Louis profited tremendously from this trade and by the end of the 1830s had drawn much of the Midwest, including Illinois, into its economic orbit. This hold was not broken until the 1850s, with the construction of the Illinois and Michigan Canal and the railroads.

Steamboat journeys were a thrilling experience, but they could be dangerous. Badly designed boilers often exploded. Fire too was a great hazard: boilers were at deck level, surrounded by stacked inflammable cargo and wooden structures. Figure 77 shows a "modern" stern-wheeler on the Mississippi River just above Cairo. Many river towns have some of these steamboats as a tourist attraction, and when several of them are gathered together, they give a faint impression of what the river must have looked like when it was alive with these colorful and noisy monsters.

79. Map of the main steamboat routes in the Northeast in the 1830s

1 Pilot House
2 Texas Roof
3 Smoke Stacks
4 Scape Pipes
5 Hurricane Deck
6 Passenger Cabins
7 Boiler Deck
8 Main Deck
9 Gang Plank
10 Cargo
11 Boilers
12 Engine Housing
13 Paddle Wheel

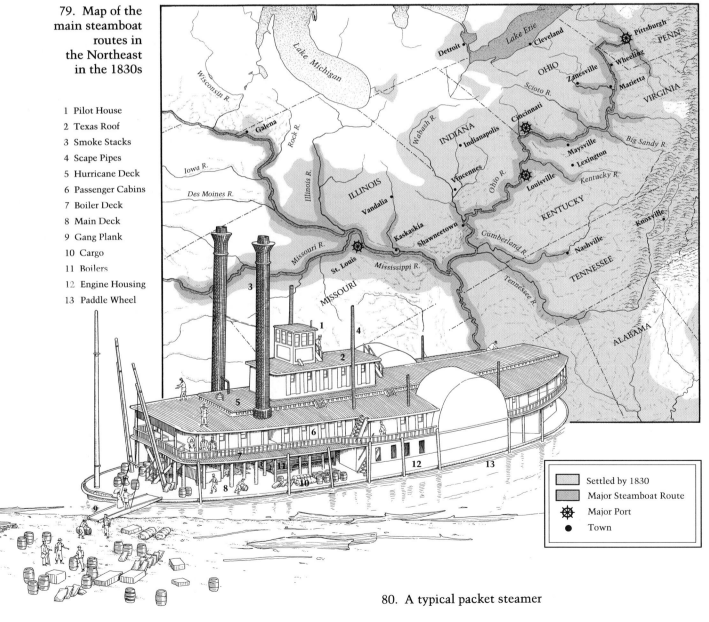

Settled by 1830
Major Steamboat Route
Major Port
Town

80. A typical packet steamer

The Northward Migration of the Capital

FROM 1800 TO 1830 ILLINOIS INCREASED greatly in population. At first this growth was confined to the eastern river valleys and to the old French sites along the Mississippi River. Towards 1820 the Mississippi, Ohio, and Wabash valleys began to fill up; Birkbeck was part of this process. During the 1820s the trend continued, so that by 1830 the settlements reached up the Mississippi almost to the Rock River, with an outpost at the Galena lead mines, and along the Illinois River and up its tributary the Sangamon River.

It is instructive to compare the maps in figures 69 and 84. The first, published in 1818, shows virtually no interior settlement north of Alton (roughly where the Illinois River joins the Mississippi). The second, published six years later, shows counties like Edgar and Sangamon beginning to fill in this blank area, and Springfield has appeared. The sequence of state capitols follows this demographic progress. In 1818 the new state was governed from Kaskaskia, using what looks like a fairly large private house (figure 81). The following year the new assembly voted to move to Vandalia, prominently marked in figure 84 in the center of the state.

At Vandalia a striking building was erected (figure 82) and served until 1839, when the capital was again moved, this time to Springfield. The Vandalia statehouse is a most evocative structure, in which the young Lincoln served as a member of the delegation from Sangamon County. Indeed, he was one of the "long nine" (tall men) who voted to move the capital to Springfield, nearer to his base at New Salem.

This move was in accordance with the changing demographic pattern in the state, and the new statehouse reflected the growing size and wealth of the population (figure 83). It is an elegant classical building, which, having in its turn been outgrown, now houses the Illinois State Historical Library. The capital had reached roughly the geographical center of the state, and in spite of the eventual growth of the Chicago region, which moved the demographic center far to the north of Springfield and led in 1865 to a proposal for a further northward migration of the capital, it has stayed in that location.

81. Photograph of the capitol building at Kaskaskia, date unknown (Illinois State Historical Library)

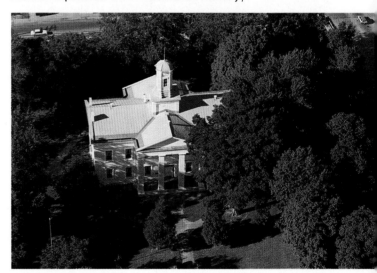

82. Aerial photograph of the capitol at Vandalia, September 1986

83. Aerial photograph of the first capitol at Springfield, date unknown (Illinois Department of Transportation)

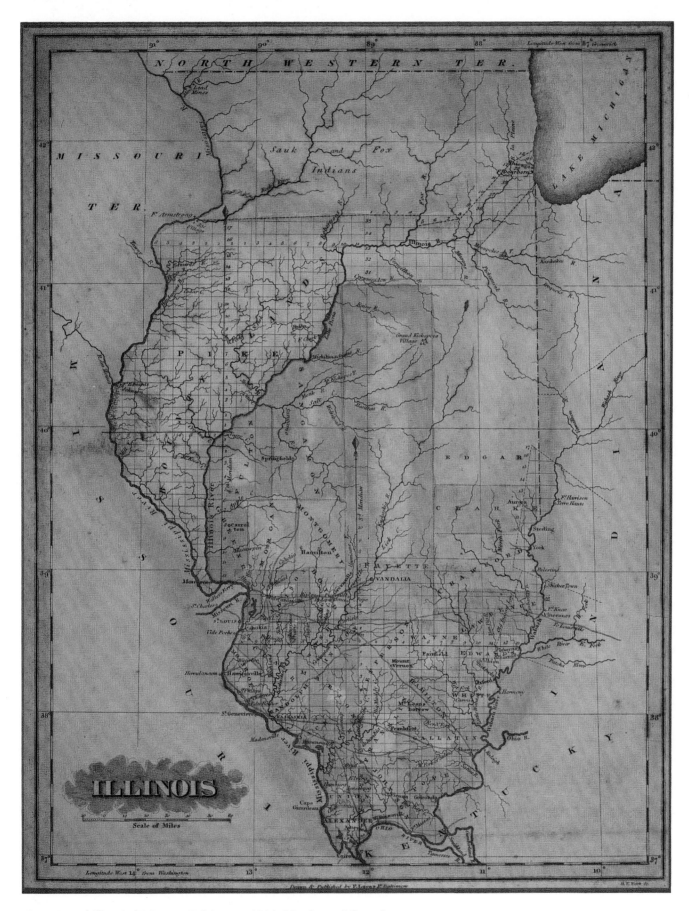

84. Map of Illinois (F. Lucas, Baltimore, 1824; Newberry Library)

5

THE DEVELOPMENT OF
THE NORTHWEST

85. Alton from the river (Henry Lewis, *Das illustrierte Mississippithal*, Düsseldorf, 1854; Newberry Library)

Introduction: Alton and the New Settlers

IN THE 1830s COMMERCE AND SETTLEMENT reached the northwestern corner of the state, a region roughly bounded by the Mississippi on the west and the Illinois and upper Rock rivers to the east. The last large Amerindian groups were forced out of the state, and a great mass of people from the East flooded in. From the late 1830s onward, the cultural and political complexion of the state changed, as Illinois began to shed its southern Anglo origins and to look more and more to the northeastern states, and even to Europe.

The town of Alton was involved in much of this early development. It stands on bluffs along the Mississippi River and has a fine natural harbor. Alton was one place where deep-draft steamboats from the lower Mississippi shifted their cargoes into shallower-draft boats suitable for the upper river. Saint Louis was the leader in this transshipment business, but in the 1820s eastern bankers identified Alton as a place that could compete for this traffic and for the growing trade of the Illinois and Missouri rivers.

The bankers encouraged merchants to establish themselves at Alton, and after the opening of the Erie Canal in 1825 many New Englanders and some Germans began to arrive and set up businesses. By the 1840s, when Henry Lewis drew his sketch of the little town, it had houses and especially churches of a markedly eastern style (figure 85). It also was novel in its general opposition to slavery, which in 1837 led to the murder there of Elijah Lovejoy by an anti-abolitionist mob. Down to the time of the Civil War, there was tension between Illinois towns like Alton and Quincy and their slave-owning neighbors across the Mississippi River.

From 1800 to 1830 the northward push of settlement, protected by strategically sited federal forts, steadily undermined the position of the Indians in the Mississippi valley. The Black Hawk Indian War, which erupted in 1831, was the culmination of this process, leading to the removal of the remaining Indian bands and their grudging resettlement west of the Mississippi River. It is possible that southerners were discouraged from settling in the northwest by the period of Indian unrest, and certain that during the 1830s they were succeeded by various kinds of settlers from the East.

Farmers and speculators came, seeking the great open lands. Merchants and bankers were eager to share in the trading of the newly developed wealth. Prospectors and miners were attracted by the mineral boom around Galena, and religious visionaries wanted to establish kingdoms of God in the new freedom of the West. Now the tide of immigration was flowing into a region of fewer rivers and more scattered timber, with vast areas of prairie. This terrain was difficult to cultivate, as Birkbeck had found out, and the relative absence of rivers made communications difficult. Slowly, a few communities such as the Swedes at Bishop Hill began to work out techniques for living on the plain, and a few centers of population began to emerge on the open prairie.

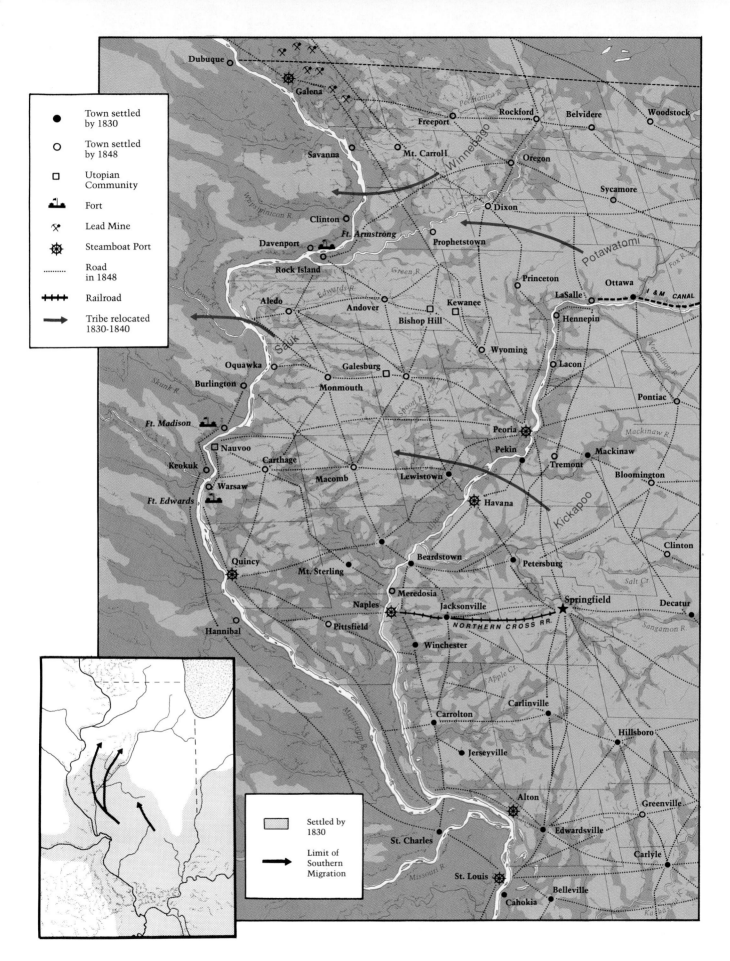

Legend

- ● Town settled by 1830
- ○ Town settled by 1848
- □ Utopian Community
- 🏴 Fort
- ⚒ Lead Mine
- ⚓ Steamboat Port
- ⋯ Road in 1848
- ┼┼┼ Railroad
- ➤ Tribe relocated 1830-1840

Dubuque
Galena
Freeport
Rockford
Belvidere
Woodstock
Savanna
Mt. Carroll
Oregon
Sycamore
Pecatonica R.
Winnebago
Rock R.
Dixon
Clinton
Ft. Armstrong
Prophetstown
Potawatomi
Davenport
Rock Island
Green R.
Princeton
Ottawa
LaSalle
I & M CANAL
Fox R.
Aledo
Andover
Kewanee
Hennepin
Edwards R.
Bishop Hill
Wyoming
Lacon
Iowa R.
Oquawka
Galesburg
Vermilion R.
Sauk
Burlington
Monmouth
Spoon R.
Pontiac
Skunk R.
Ft. Madison
Peoria
Mackinaw R.
Nauvoo
Pekin
Mackinaw
Keokuk
Carthage
Macomb
Lewistown
Tremont
Bloomington
Warsaw
Havana
Kickapoo
Ft. Edwards
Clinton
Beardstown
Quincy
Petersburg
Salt Cr.
Mt. Sterling
Springfield
Decatur
Naples
Jacksonville
NORTHERN CROSS R.R.
Sangamon R.
Hannibal
Pittsfield
Meredosia
Winchester
Illinois R.
Apple Cr.
Carlinville
Hillsboro
Carrolton
Jerseyville
Greenville
Mississippi R.
Alton
St. Charles
Edwardsville
Carlyle
Missouri R.
St. Louis
Belleville
Cahokia
Kaskaskia R.

Settled by 1830

Limit of Southern Migration

86. Map showing northwestern Illinois
from 1830 to 1848

87. Fort Armstrong from the Mississippi River (Henry Lewis, *Das illustrierte Mississippithal*, Düsseldorf, 1854; Newberry Library)

The Military Outpost at Rock Island

Duri ng the War of 1812, the Sauk Indians under Black Hawk were the allies of the British and more than held their own against the federal forces. At the end of the war, therefore, it was decided in Washington that a line of forts had to be established on the Mississippi River, to enforce federal claims. One of these forts was to be at the mouth of the Rock River, close by the principal Sauk village; at the same time, a large tract of land was set aside for veterans in order to promote settlement of the region.

The site chosen for the fort was on Rock Island, in the Mississippi River at the mouth of the Rock River; there in 1816 Fort Armstrong was built. As figure 87 shows, it did not look much like earlier forts such as Fort Crevecoeur (figure 34) or Fort de Chartres (figure 52). Its key component was the blockhouse, a substantial square fort placed at the corners of the stockade, in which the garrison could hope to hold out indefinitely against attack by Indians; it could not withstand artillery fire. At Fort Armstrong three of these blockhouses were linked by other buildings and by a stockade to form a fortified enclosure.

Figure 89 is a modern view of Rock Island from the south; it gives an excellent idea of the strategic placement of Fort Armstrong (indicated by a box) on its low bluffs. We have to imagine away the V-shaped dikes at the island's southern tip and can then see the site as it was in the early nineteenth century.

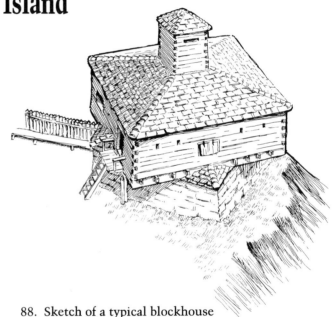

88. Sketch of a typical blockhouse

89. (opposite) Aerial photograph of Rock Island, 1959 (Department of the Army)

When Henry Lewis drew the fort for his work on the Mississippi valley (figure 87), he noted that it was made of red cedar logs and covered both branches of the river. It ceased to be an active fort in 1836, becoming a storage depot, and was largely destroyed by fire in 1855. On the site now is a reconstructed blockhouse, and on the rest of the island is the immense arsenal described in chapter 11.

90. Aerial photograph of the Black Hawk statue at Oregon, October 1988

Black Hawk and the Expulsion of the Indians

91. Portrait of Black Hawk (Illinois State Historical Society Library)

DURING THE EARLY PERIOD of European penetration of the Illinois country, relations with the Indians were relatively friendly. The French around Kaskaskia and Cahokia were too few seriously to alarm their Indian neighbors, and the occupation of the southern regions of Illinois (described in chapter 4) involved lands where Indians were not thickly settled.

It was very different in the area being flooded with settlers in the 1830s. There along the Rock River lay some of the land most prized by Indians for its fertility, and there were the great villages of the Sauk and Fox, who at this time numbered about five thousand people. Fort Armstrong itself was built on the island that Black Hawk described as "our garden (like the white people have near their big villages), which supplied us with strawberries, blackberries, plums, apples and nuts of various kinds; and its waters supplied us with pure fish, being situated in the rapids of the river."

Some Indians, led by Keokuk, argued that in the face of overwhelming numbers they should accept the offers of the federal government and resettle west of the Mississippi River. Others, led by Black Hawk, felt the impending loss too keenly to leave without a fight. In 1832 matters came to a head

92. Indians using travois (Henry Schoolcraft, *The Indian Tribes of the United States*, Philadelphia, 1851–57; Newberry Library)

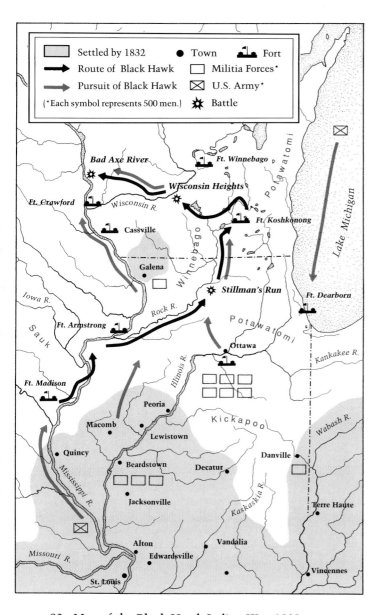

93. Map of the Black Hawk Indian War, 1832

when Black Hawk, who had been trying in vain to subsist on the lands assigned to him across the Mississippi River, crossed over to Illinois and led a band up from the south and along the valley of the Rock River. This band was probably not intent on war, as it included women and children; Black Hawk's aim was to go and plant in his ancestral lands, since his people were starving in their new lands across the river.

The panic of the new settlers led to a number of skirmishes, one of which ended badly for the militia. More of the militia were called up—among them a certain Abraham Lincoln of New Salem—and Black Hawk's band was pursued into Wisconsin. Eventually the militia and regular soldiers drove them westwards to the Mississippi, where most were slaughtered at the so-called Battle of Bad Axe, as they were trying to cross the river. Black Hawk survived another six years, long enough to see a great influx of settlers in his beloved valley.

He is commemorated in Illinois by various monuments, of which the finest is the huge cement statue on the bluffs at Oregon, constructed by Lorado Taft in 1910 (figure 90). At this site between 1898 and 1936 Taft was the animator of an artists' colony known as Eagle's Nest; today one of the cottages that he built houses the faculty offices of Northern Illinois University's Lorado Taft Field Campus.

Some of the settlers in the Rock River valley were former soldiers who had observed the valley's great promise during the 1832 campaign, but others came from the East, using the immigration routes along the Erie Canal, across the lakes, and through the port of Chicago, as well as the National Road. The northern parts of Illinois would be settled mostly by migration from the East, giving Illinois a political complexion very different from what it had been.

94. View of Galena from the north about 1855 (lithograph by Endicott and Co., New York, n.d.; Illinois State Museum)

The Development of Lead Mining around Galena

NOT ONLY THE RICH FARMLAND drew many easterners to northwestern Illinois after the defeat of Black Hawk. It had been known for many years that lead could be found along the middle Mississippi River. The Indians mined it north and south of the mouth of the Wisconsin River, and the French were active in this area from the 1690s onwards. All this was on a small scale, however, and often for local consumption.

As the republic expanded westwards, new markets opened up, with vastly improved methods of communication to serve them. During the 1810s lead was mined in a small way on the Fever River and shipped in flatboats down the Mississippi. But the mining industry really took off after 1822, when the federal government leased a tract to Colonel James Johnson, who brought twenty miners and fifty slaves from Kentucky to work his claim. In the following year the *Virginia* became the first steamboat to climb the upper Mississippi and reach "the point," as the site of Galena was known until 1826, and the elements were in place for rapid expansion.

The town of Galena, laid out in 1826, became the capital of a great lead-producing region. In the mineral-rich hills of the driftless area a multitude of small mines developed, using the kind of elemen-

95. A lead mine about 1850 (David Dale Owen's report on the Wisconsin lead mines, 28th Cong., 1st Sess., Senate Document 407; Newberry Library)

96. Aerial photograph of Galena from the
south, April 1988

tary techniques illustrated in figure 95. Production
increased during the 1830s and 1840s, until in 1845
a record fifty-four million pounds of lead was
shipped out of Galena, which had become one of the
largest towns in Illinois and was the largest supplier
of lead in the United States. Figure 94, a view from
the north made in the middle of the nineteenth century, shows the town at the height of its prosperity,
when as many as seven steamboats might simultaneously be maneuvering in the river.

Thereafter production slowly declined, as the eas-
iest seams played out and cheaper sources were
found in Missouri and the West. Wholesale clearing
of timbered hills led to serious erosion, and eventually the river silted up to such a degree that steamboats could no longer use it. Galena entered a period
of somnolence as a well-preserved mid-nineteenth-
century town, from which it has recently emerged
as a major tourism center.

Figure 96 shows it from the south, with the Galena River (formerly the Fever River) winding across
the photograph; note how few intrusions there are
into the pattern of nineteenth-century houses. Except in floodtimes the river is now quite small, but

97. Detail of the view of Galena

figures 94 and 97 show us how it looked in the days
when steamboats could dock alongside the levee to
unload all the supplies necessary for the burgeoning
town and reload with the lead pigs.

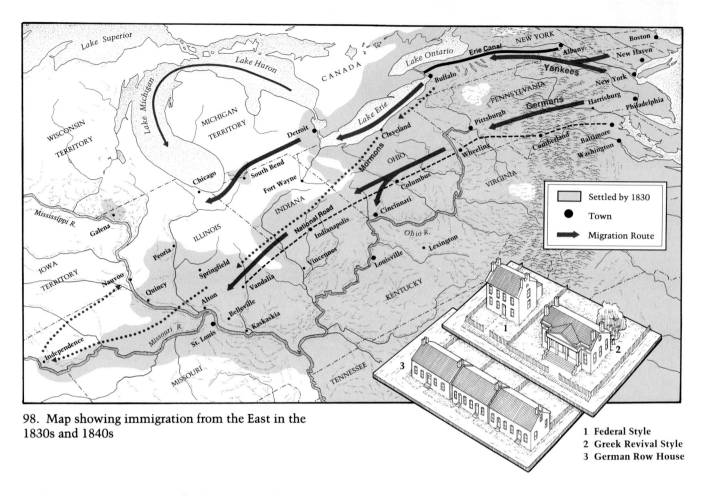

98. Map showing immigration from the East in the
1830s and 1840s

1 Federal Style
2 Greek Revival Style
3 German Row House

The Coming of the Yankees

EASTERN MIGRATION TO NORTHWEST ILLINOIS
continued unabated until the great depression of
1837 slowed its pace, though only temporarily;
by 1840 the population of the state had reached
450,000. Yankees came via the Erie Canal route, settling throughout the northwest, and Germans came
via the National Road, establishing themselves primarily in Belleville and in Madison and Saint Clair
counties, though many settled in towns like Alton
and Peoria. The new people quickly made their influence felt in commerce, politics, and culture.

New kinds of architecture and town plans reflected the new cultures. The easterners built brick
or frame Federal-style Greek Revival homes and surrounded them with white picket fences and decora-
tive gardens. Germans often built brick row houses,
facing directly onto the street. In commercial terms,
towns like Quincy and Jacksonville marked the
growing power of the easterners. Quincy, located on
the upper Mississippi between Rock Island and Alton, enjoys an admirable site, with bluffs overlooking a convenient mooring place. As its name suggests, it dates back to the presidency of John Quincy

Adams, having been platted in 1825, and it became
an early center for German and eastern immigration.

When Henry Lewis drew the little town about
1848 (figure 99), it was in a period of rapid growth,
based on the river with its steamboats, on northern
capital (of which we see a hint in the Federal-style
buildings of the waterfront), and on the fertile hinterland. Quincy used these advantages to develop a
wide range of commercial and industrial activities.
It was a great exporter of hogs, a great center for milling corn, and a considerable producer of tobacco. It
had breweries, paper mills, distilleries, timberyards,
brickworks, printing houses, and iron foundries. As
time went by more and more Germans came into
the region, and by the 1850s Quincy was the second
largest town in Illinois, a position that it held until
about 1870.

Yankee immigrants founded many schools and
colleges, some of which did not survive. Jubilee College was established a little to the west of Peoria in
1838 by Bishop Philander Chase. Born in New
Hampshire in 1775, he was ordained an Episcopal

priest in 1798 and thereafter lived in many parts of the country, always establishing churches and schools just behind the advancing frontier. After working in Louisiana and Indiana he went to Ohio, where he set up Kenyon College. No doubt he hoped that Jubilee College would be as successful as Kenyon, but a combination of poor management and lack of pupils led, after his death, to its slow decline. The buildings have been restored and give an excellent idea of the type of vision possessed by the Yankees of the early nineteenth century (figure 100).

99. Quincy from the river (Henry Lewis, *Das illustrierte Mississippithal*, Düsseldorf, 1854; Newberry Library)

100. Engraving of Jubilee College, nineteenth century (Laura Chase Smith, *The Life of Philander Chase*, New York, 1903; Newberry Library)

101. Aerial photograph of Bishop Hill, September 1988

Bishop Hill and the Early Settlement of the Prairies

MOST OF THE IMMIGRANTS to northwestern Illinois came alone or in family groups, but a few communities immigrated en masse. The Mormons for a time settled at Nauvoo, and a group of Swedes, led by Eric Jansson, in 1846 settled on the prairie at Bishop Hill. This colony prospered for a brief period, though Jansson himself was murdered in 1850, and it has left a distinctive town.

By the late 1830s some farmers in northern Illinois were beginning to experiment with crops on the prairies and discovered that the grasslands were not as barren as had been thought. But individual farmers as yet lacked the tools to break the thick prairie sod. A colony like Bishop Hill could bring to agriculture the disciplined labor of a large number of men and women, and the Swedes showed that the prairie could be tamed by sheer force of manual effort, even before the advent of improved plows and mechanical reapers. Our sequence of three figures shows members of the colony breaking the sod, sowing corn and reaping wheat.

For breaking the prairie with the traditional iron-tipped wooden plow, a six-ox team was necessary but was outside the means of most individual farmers. Planting corn was a communal effort, with the women lined up behind a cord, inserting the seeds as the supervisors slowly advanced the cord across the field. This was a much less laborious way of planting corn than the interminable walking and stooping of the individual farmer. In the reaping stage the men worked into the ripe wheat, while the women bound and stacked the sheaves behind them. These images of planting and reaping powerfully evoke a system capable of substantially increased production. However, the great wealth of the prairies was eventually released by individual farmers, using a very different system of production.

103. Olof Krans, *Breaking Prairie* (Bishop Hill State Historic Site, Illinois Historic Preservation Agency)

104. Olof Krans, *Corn Planting* (Bishop Hill State Historic Site, Illinois Historic Preservation Agency)

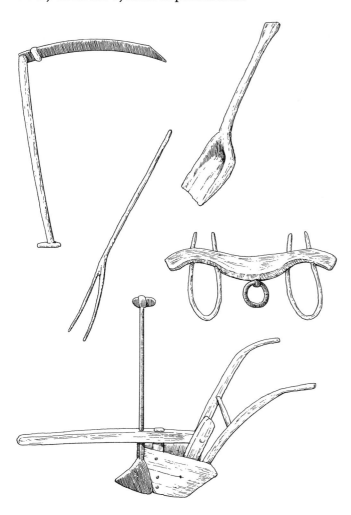

102. Farm implements of the early 1800s

105. Olof Krans, *Harvesting* (Bishop Hill State Historic Site, Illinois Historic Preservation Agency)

106. Aerial photograph of Peoria, about 1940 (University Library, Champaign-Urbana)

Peoria and the Development of the Illinois River Valley

Peoria, delightfully situated on a lake in the Illinois River, was a center of Franco-Indian settlement in the late seventeenth century. Occupation of the site seems to have been somewhat sporadic during the eighteenth century, but the town was platted in 1826, and two years later the first steamboat reached the area. In 1835 the town was incorporated and began to grow steadily, especially as a center for distilleries.

Its plan is unusual (figure 106), for onto the existing French street plan, with its northwest-southeast orientation to the lakefront (though without long-lots), was grafted a new section conforming to the north-south and east-west lines of the General Land Office survey. In this case the older alignment was much more functional, for its streets conformed to several bluff lines, parallel to the shoreline, which give Peoria its distinctive geographical character.

Lying near the upper limit of steamboat navigation on the Illinois River, Peoria became during the 1830s and 1840s a major port in the river's economy (figure 108). It was the chief distribution point for the produce of the rich farmlands of the Illinois River valley, sending great quantities of agricultural products down the river to Alton, and from there to Saint Louis and even to New Orleans. Rich clay de-

107. Charles Peck, Peoria about 1858 (lithograph, Reen and Shober, Chicago, 1858; Chicago Historical Society)

Legend:
- ● City or Town
- ⊛ Important Steamboat Port
- ┼┼┼ Railroad
- --- Projected Canal to Lake Michigan
- ▭ Emerging Corn Belt
- ⬛ Pottery Manufacture

108. Map of the Illinois River in the 1840s

posits were found along the southern half of the Illinois River, and many of the little towns along the river developed potteries, whose wares were highly prized throughout the region.

During the 1830s Peoria enjoyed growing prosperity, but her merchants and traders, like many others in northern Illinois, were more and more dissatisfied with their ties to the southern markets. The major demand at this time lay in the East, to which access for goods was relatively difficult. In 1836 work began on the Illinois and Michigan Canal (chapter 6). When the canal was opened in 1848, produce from the lower Illinois River valley could be moved cheaply to Chicago, the Great Lakes, and beyond, and eastern and foreign products flooded into the Peoria region. Peoria became in a sense the western terminus of the canal and profited commensurately. How the town looked in 1858 is shown in figure 107; steamboats filled the lake, and to the left are the smokestacks from numerous agricultural industries.

6

THE EMERGENCE
OF CHICAGO

The Indian Boundary Lines and Federal Planning

IN THE 1670s JOLIET REALIZED that the portage at "Checagou" between the Great Lakes and the Mississippi River was likely to become a fulcrum of economic and political power. This dream was not immediately realized, as Indian hostility made the portage perilous for much of the eighteenth century. Some traders did get through, but most seem to have used the Kankakee River route to the south.

With the coming of independence, statesmen in Washington also grasped the strategic importance of the area and saw that the republic needed to control this crucial watershed. In 1796 the Treaty of Green-ville provided for a military reservation at the mouth of the Chicago River, and in 1803 the first Fort Dearborn was constructed there. After the War of 1812 (marked at Chicago by the Dearborn massacre and destruction of the fort), the United Tribes (the Sauk, Fox, and Potawatomi) ceded to the federal government a corridor of land from Lake Michigan to the Illinois River within which to build a canal.

Figure 109 shows this corridor and marks the emergence of the town of Chicago, lying at its north-eastern end. The diagonal Indian Boundary Lines de-fine the corridor to the north and to the south, with-

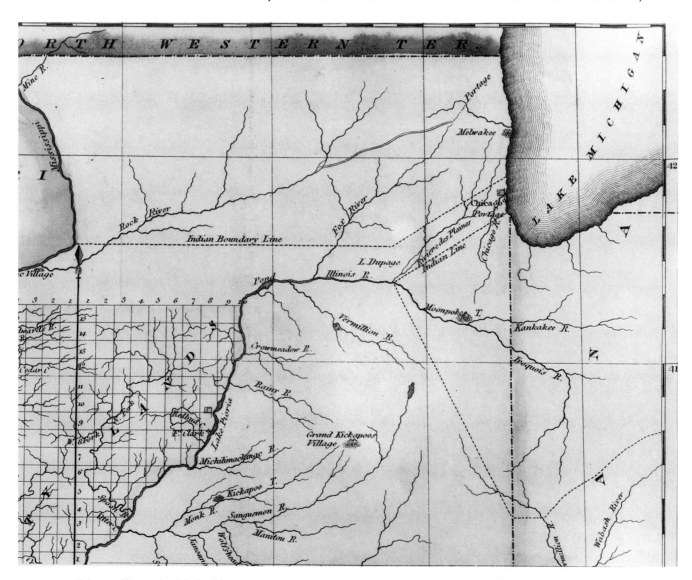

109. Detail from *Illinois* (published by H. C. Carey and I. Lea, Philadelphia, 1822; Newberry Library)

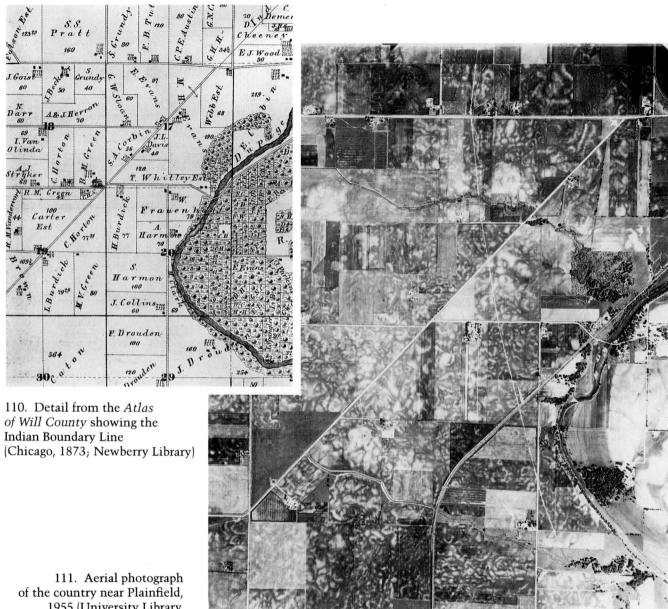

110. Detail from the *Atlas of Will County* showing the Indian Boundary Line (Chicago, 1873; Newberry Library)

111. Aerial photograph of the country near Plainfield, 1955 (University Library, Champaign-Urbana)

out making contact with the township-and-range system, which was still working up from the southwest. Note that at this time the land to the north and northwest was still a territory.

The Indian Boundary Lines survive on many maps down to the present and are in places fixed in the street pattern. The southern line has become a short stretch of I-57, where it goes through Markham. The eastern end of the northern line became Rogers Avenue in the far northeast of the city, while for a stretch by River Grove it is called Forest Preserve Drive. In the southwest, after disappearing for many miles, this northern line suddenly emerges as a diagonal road in the middle of the fields southwest of Plainfield. It is instructive to compare the aerial photograph of this road (figure 111) with the 1873

Atlas of Will County (figure 110). There is no apparent reason for the sudden emergence of this diagonal, which in the middle of the map and of the aerial photograph causes a marked disjuncture in one east-west road. The area to the east of the Du Page River is much less heavily wooded in 1955 than it was in 1873, but in other respects there is considerable continuity, even to the outline of some of the farms.

In the very middle of the photograph the main east-west road makes a little jog southwards where it meets the Indian Boundary Line; this jog has survived from 1873. About the only major novelty in the aerial photograph is the railroad track coming in from the right and exiting at the center bottom; these are the lines of the Elgin, Joliet, and Eastern Railroad.

112. Henry Schoolcraft, *View of Chicago*, 1820 (*Chicago Magazine*, 1857; Chicago Historical Society)

Chicago in 1829

CHICAGO IN THE 1820S GAVE LITTLE INDICATION of future greatness. Figure 114 shows what this backwater outpost might have looked like in 1829, just before it was platted as a canal town. The Chicago River is the dominant feature, with its North and South branches. The portage to the Illinois River was reached by the South Branch and in 1829 had been in use by travellers and fur traders for many years. At most seasons the river was very sluggish, so that where it entered the lake the southerly lake current formed a huge sandbar, which the river was quite unable to overcome. Figure 112 shows how the settlement looked from the lake off this sandbar, which made navigation of the river impossible except for very light-draft vessels like canoes.

Fort Dearborn, which was rebuilt after its destruction in 1816, lay just by the south end of the future Michigan Avenue Bridge. The fort had its own vegetable garden and a barn to quarter its horses; its single blockhouse was well positioned to cover the landward approaches. Just to the south of the fort lay the government factor's house, and to the south of that was the trading establishment of Jean Beaubien, agent for John Jacob Astor's American Fur Company. Then came the graveyard and parade ground, alongside the old Vincennes road. This ran for some way along the beach, eventually reaching the old French settlement on the Wabash (see chapter 3).

The road to the north was the Little Fort Road, which left from the spot where Kinzie (and once Jean-Baptiste Point du Sable) had his trading station, and eventually reached Green Bay in what is now Wisconsin. The woods were quite dense on this north side but did not contain much timber suitable for building. Wolf Point (figure 113) was the other focus of settlement, with a bridge, a ferry, and two

113. George Davies, *Wolf Point*, in 1834 (*Chicago Magazine*, 1857; Chicago Historical Society)

taverns; these served the river trade and travellers setting out on the many trails to the northwest. Mark Beaubien's Sauganash Inn was noted throughout the region for its owner's hospitality, even if its comforts were very simple.

"Checagou" was still very thinly settled, and indeed the whole area was probably no more active than it had been fifty years earlier, when the Indians were more numerous and the fur trade still prospered. The year 1829 marked a low point in the history of the little settlement, for the opening of the Erie Canal in 1825 had yet to have much effect, and the fur trade was dying off. It would have been a bold person who forecast even a modest prosperity for this backwater, much of which was under water at certain times of the year.

The conventional assessment of Chicago's potential was given by Professor William Keating, who in 1823 accompanied a government survey to the region as mineralogist. "The communication will be limited, the dangers of navigation on the lake, the scarcity of harbors, must ever prove a serious obstacle to the increase of the commercial importance of Chicago. Indeed, when the game is gone, it is doubtful that even the Indians will reside here much longer."

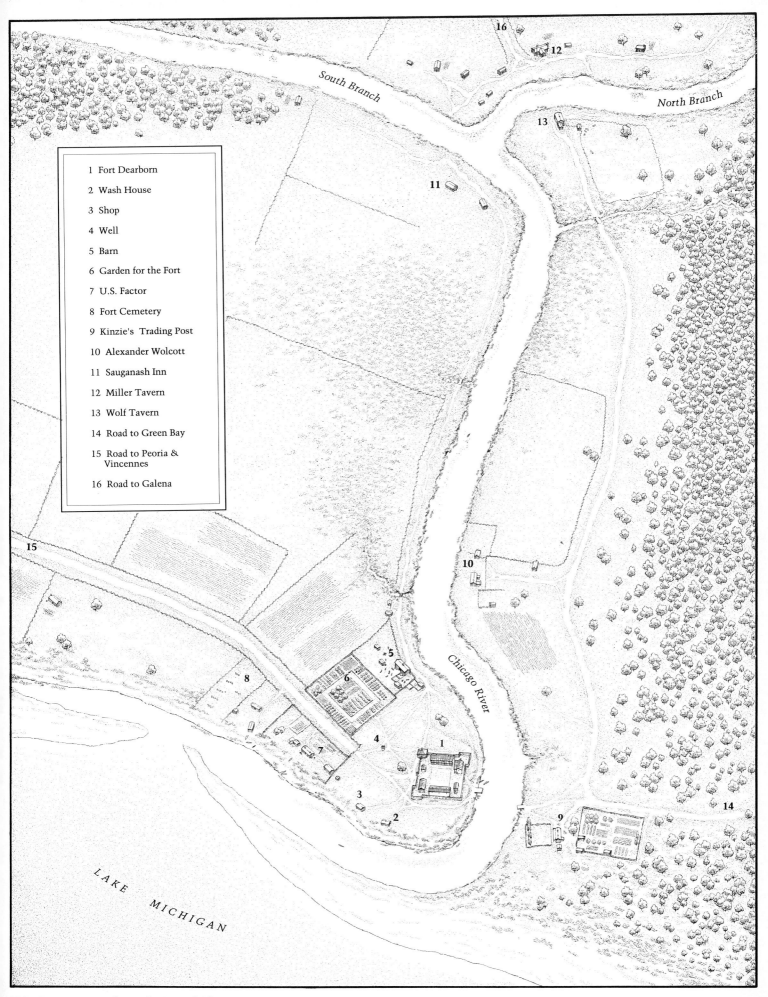

1 Fort Dearborn

2 Wash House

3 Shop

4 Well

5 Barn

6 Garden for the Fort

7 U.S. Factor

8 Fort Cemetery

9 Kinzie's Trading Post

10 Alexander Wolcott

11 Sauganash Inn

12 Miller Tavern

13 Wolf Tavern

14 Road to Green Bay

15 Road to Peoria & Vincennes

16 Road to Galena

South Branch

North Branch

Chicago River

LAKE MICHIGAN

114. Reconstructed aerial view of Chicago in 1828

The Transport Revolution on the Great Lakes

115. Aerial photograph of the *Charlotte Anne*, about 1985 (by courtesy of Dave Pasquith)

THE GREAT LAKES SAW SAILING SHIPS before the nineteenth century, for La Salle had built and lost the *Griffon* there, and the United States and British navies had sailed their small warships throughout the lakes. But the completion of the Erie (1825) and Welland (1829) canals meant that this enormous area of landlocked water was open to the Atlantic Ocean and to the East Coast of the United States, thus creating possibilities for immigration and trade between the agricultural regions of the Midwest and the growing cities of the East.

By this time the Great Lakes had upwards of 150 sailing vessels and around 15 steamships in operation. Lake Erie had most of this traffic, followed by Lake Ontario. Lakes Huron and Michigan still had very little, and Lake Superior none at all. All that would change in the next few years, as the westward flow of immigrants began to penetrate the whole area.

The influx began about 1834, and in 1835, 255 sailing ships arrived in Chicago, a figure that increased each year, as the new settlers sent back word of the vast agricultural lands still available. During the 1830s and most of the 1840s, the trade in agricultural produce came mostly out of Ohio and Indiana, whether to Lake Erie or down the Ohio River and on to New Orleans. But when the Illinois and Michigan Canal was finished in 1848, western fertile lands added their quota to the eastward flow.

On rivers and lakes the steamboats were indispensable, for they could maneuver without regard for wind or current. However, they were not well suited to carrying large bulk products such as grain or coal, and between 1830 and 1890 much of this cargo was carried by sailing vessels. Their construction called for no expensive materials or novel engineering skills, and many yards around the lakes could build them.

The lake boats varied widely in type. Some were little more than sailed barges, designed to fit the Welland Canal. Others tried to take into account the peculiar sailing conditions of the lakes and began to develop novel features. One of the most successful was the clipper-type schooner developed at Manitowoc in the 1850s (figure 117). It usually had two masts and a relatively flat bottom; stability was ob-

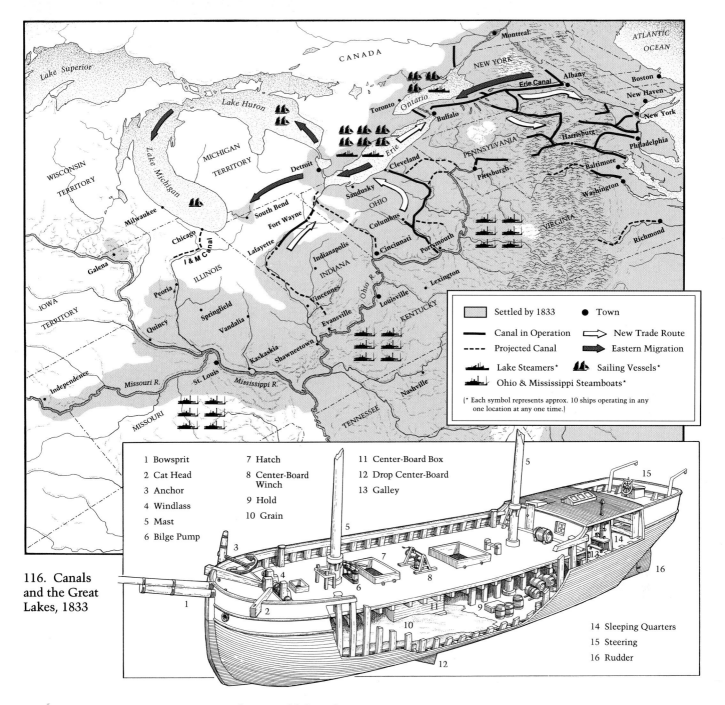

116. Canals and the Great Lakes, 1833

Legend:

Settled by 1833 ☐   ● Town
Canal in Operation ──   ⇨ New Trade Route
Projected Canal ----   ⬛ Eastern Migration
Lake Steamers*   Sailing Vessels*
Ohio & Mississippi Steamboats*

(* Each symbol represents approx. 10 ships operating in any one location at any one time.)

1 Bowsprit
2 Cat Head
3 Anchor
4 Windlass
5 Mast
6 Bilge Pump
7 Hatch
8 Center-Board Winch
9 Hold
10 Grain
11 Center-Board Box
12 Drop Center-Board
13 Galley
14 Sleeping Quarters
15 Steering
16 Rudder

117. Cutaway view of a typical lake schooner

tained with a centerboard, in effect a movable keel.

As time went by the schooners grew not only in number but also in size and complexity. They tended to be longer and narrower than saltwater ships but could when necessary hold their own upon the world's oceans; some crossed the Atlantic Ocean and others rounded the Horn. They often specialized in particular trades; on Lake Michigan, for instance, some brought timber from Michigan for Chicago's lumberyards, and others brought northern ores that, when combined with coal from northern Illinois, fed the great iron foundries at the south end of the lake. One way and another, the schooners were crucial to the development of Illinois in the decades before the Civil War and continued to be important even after that. Very few survive; figure 115 shows the *Charlotte Anne*, a schooner that still takes passengers on trips out of Chicago.

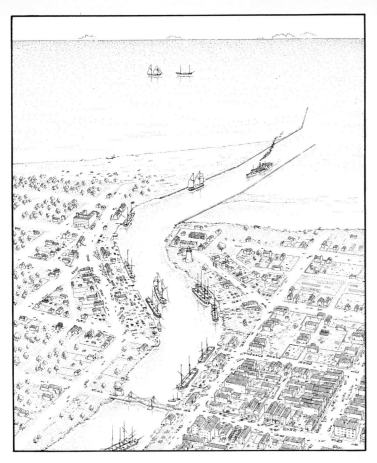

118. Reconstructed view of the mouth of the Chicago River in 1839

119. Aerial photograph of the mouth of the Chicago River, about 1970 (Chicago Association of Commerce and Industry)

120. The *Osceola* loading grain in 1839 (Newberry Library)

The Portage Becomes a Port

CHICAGO IN 1829 was not a promising site for a city. In the years that followed, however, tremendous pressures developed for the establishment of a port there. It was the designated eastern terminus for the proposed canal, and the natural point of entry for the flood of goods and immigrants from the East, since most of the western shore of Lake Michigan lacks good natural harbors. In the early 1830s Chicago began to be recognized as a coming town, especially after it was platted in 1830 and a large number of land speculators and immigrants began to arrive. By 1837 the population had reached four thousand.

Sailing vessels and the occasional steamship coming to Chicago had to lie off the sandbar and laboriously ferry their passengers and cargo ashore. Clearly the harbor mouth had to be improved, and in 1833 work began, using federal funds for the construction of two piers that would cut through the sandbar and offer an unobstructed passage from the lake into the river. By 1834 vessels could enter the harbor between the growing piers, and after the northerly one had been extended to deal with sand moving south in the current, the harbor looked rather as it does in figure 118.

Two schooners, perhaps loaded with Michigan pine or immigrants, are lying off the harbor mouth, waiting to come in. Another is sailing out between the piers. Fort Dearborn is clearly recognizable, but all around it are hurriedly built houses and stores, and a line of wharves was emerging along both sides of the river. Figure 120 shows the *Osceola* loading grain in 1839 at one of these wharves near the river mouth; this was the beginning of what would become a huge export trade. On the south bank, a business district was beginning to develop at Lake and South Water streets.

Just by Fort Dearborn was a lighthouse, one of the many that were built to make the lake moderately safe for navigation. Even after a series of lighthouses was established, the lake continued to be dangerous,

for it is subject to sudden fierce storms and ships often have little seaway and few harbors in which to escape them. Figure 121 is an aerial view of the lighthouse at Gross Point, built in 1870–73 after wrecks like that of the *Lady Elgin*, in which three hundred people lost their lives just off Winnetka.

It is hard now in downtown Chicago to peel the layers back and imagine how the harbor looked in 1840. But figure 119 reveals surprising elements of continuity. Not only is the rather distinctive curve of the river the same, but the pronounced northward bend at the tip of the northerly pier may still be seen, as part of the riverbank just west of the Lake Shore Drive Bridge. The land now extends far out into the lake, in part as a result of the build-up of sand to the north of the northerly pier.

121. Aerial photograph of the lighthouse at Gross Point, June 1985

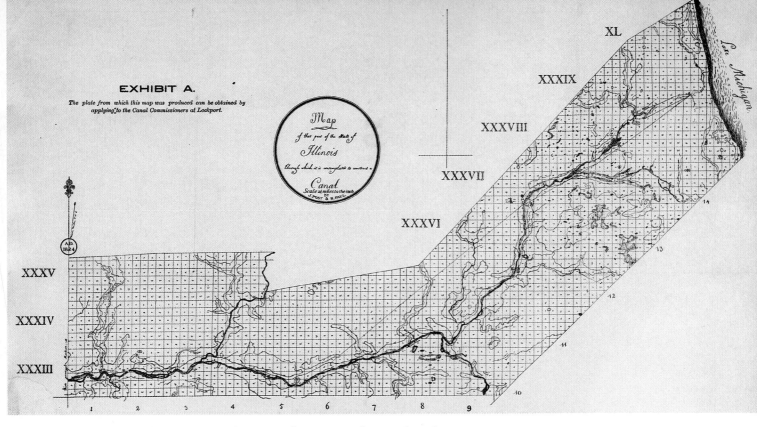

EXHIBIT A.

The plate from which this map was produced can be obtained by applying to the Canal Commissioners at Lockport.

122. Canal map (Justus Post and René Paul, 1823; Newberry Library)

Work Begins on the Canal

THE ERIE CANAL, as well as its European precursors, demonstrated the extraordinary influence that such works could have on economic expansion. For at least two decades before work on the Illinois and Michigan Canal began in 1836, visionaries had pressed for its construction. By the early 1830s the time was ripe; the Black Hawk Indian War was over, Chicago was emerging as a port, and settlers were moving into the agricultural area that the canal could serve.

Headquarters for the project (the building with a pillared veranda in figure 123) was constructed in the newly platted town of Lockport; it is today the Will County Historical Society Illinois and Michigan Canal Museum and contains interesting exhibits of those early days. Just behind it and to the right is the Gaylord Building (now extensively restored and housing exhibits), which was built to store the construction equipment, and under the trees beyond that is a section of the canal.

The land within the Indian Boundary Lines was platted into the usual townships and sections (figure 122), and the canal commissioners were assigned 284,000 acres by the federal charter grant of land made in 1822. The commissioners raised money for construction by selling off certain sections; their

Lockport headquarters served as a land office as well as an engineering center.

Much of the work was arduous, for it involved digging through layers of limestone, using only hand tools and black powder. The most difficult part of the work, technically speaking, was the construction of the locks, shown in figure 125. After the channel had been dug out, stone was brought on

123. Aerial photograph of the canal headquarters, October 1985

124. Aerial photograph
of Saint James of the Sag,
April 1986

125. Reconstructed view of the
building of a lock

carts from nearby quarries and, by using only sheerlegs, was precisely positioned to form the locks that
we see today. It took a great deal of engineering skill
to lay out the line of the lock and to construct it to
hold the wooden doors snugly, preventing undue
leakage.

The men directing the enterprise came from
North America and Europe, but the hard digging was
done mostly by Irish immigrants, many of whom
were hired just off the docks in New York. They
formed their own community near the northern end
of the canal at Bridgeport, which has been influential in the life and politics of Chicago ever since. The
work was physically very demanding, and as the
marshes and wet prairies of the valley had endemic
fevers, particularly during the summer, mortality
was high. Many of the laborers rest in the cemetery
of Saint James of the Sag, beautifully sited on a little
knoll overlooking the valley of the Des Plaines River
(figure 124).

Chicago's First Boom and Bust

THE OPENING OF THE RIVER MOUTH, the beginning of the canal, and the continuing immigration into the region combined to set off a great land boom in Chicago from about 1832 until the crash of 1837. At first houses in the new town were built in the time-honored way, using heavy custom-made beams and posts to produce such elegant structures as the Widow Clarke House (figure 127). Built in 1836, the house was moved twice before reaching its present site in 1981 and gives us a good idea of the appearance of the Greek Revival houses of the period.

The construction methods used were too time-consuming for the enormous demand for houses, however, and a much cheaper and speedier method was soon developed, the balloon frame (figure 128). Presawed planks were nailed together to form a frame, which was then covered with shingles or tar over a wooden roof; critics claimed that the resulting "balloon" would be vulnerable to heavy weather, but

in fact it was surprisingly strong, and many houses are still built this way. The whole structure stood on a foundation of stone or brick, which tended to sink into the Chicago mud. It was not very elegant, but it was a practical solution to an urgent architectural problem and made the best use of the pine that was being shipped in huge quantities from Michigan and Wisconsin.

The frenzied expansion of the period 1832–37 gave way to a deep depression, which lasted into the early 1840s. It led for a time to the suspension of work on the canal and to a much diminished rate of building in Chicago. Still, by 1845 the straggling settlement of 1830 had become a substantial little town. Figure 126 shows it from the west in 1845. In the foreground cows are grazing on the prairie. Then we see some of the hastily constructed houses and behind them, among the ships' masts, the towers and spires of a number of churches. The bust was over, and growth was beginning again.

126. Engraving of Chicago from the west in 1845 (Norris City Directory, 1845; Newberry Library)

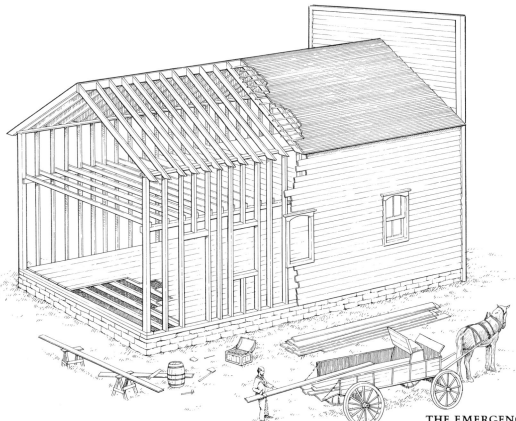

128. Drawing of balloon-
frame construction

129. Aerial perspective of the length of the Illinois and Michigan Canal in 1848

Profile of the Canal: Chicago to La Salle

WHEN THE CANAL OPENED in 1848, it linked Chicago with La Salle, where boats could continue down the Illinois River to the Mississippi River. The canal was at least six feet deep and sixty feet wide; to allow for the 115-foot drop in level from Chicago to La Salle, it had fifteen locks. It also had two aqueducts, to cross Aux Sable Creek and the Fox River.

Figure 129 shows the length of the canal, as it was about 1848. At Chicago the canal left the south branch of the Chicago River and ran southwest through what had been the mud lake, on the line of the old portage. Then for miles it ran alongside the Des Plaines River, which at its junction with the Kankakee River becomes the Illinois River. It followed this westwards to La Salle, where a large turning basin allowed boats to pass into the Illinois River.

As the perspective shows, the canal ran much of its western course beneath bluffs, which had been formed after the last ice age (see chapter 1). The canal had the disadvantages of freezing for about three months of the winter and of tending to run low during the dry summer months, in spite of the water supplied by four feeder canals.

The early barges were of two types, shown in the figure 129 inset. The upper one is a freight barge, and the lower a passenger packet; both were built to fit the fifteen locks, which were roughly 18 feet wide by 110 feet long. The freight barges were usually towed by mules and could pass out into the Mississippi River or into Lake Michigan, where they were towed by steamboats in a linked train. The passenger packets often used horses to tow them; these were faster than mules and made the journey from

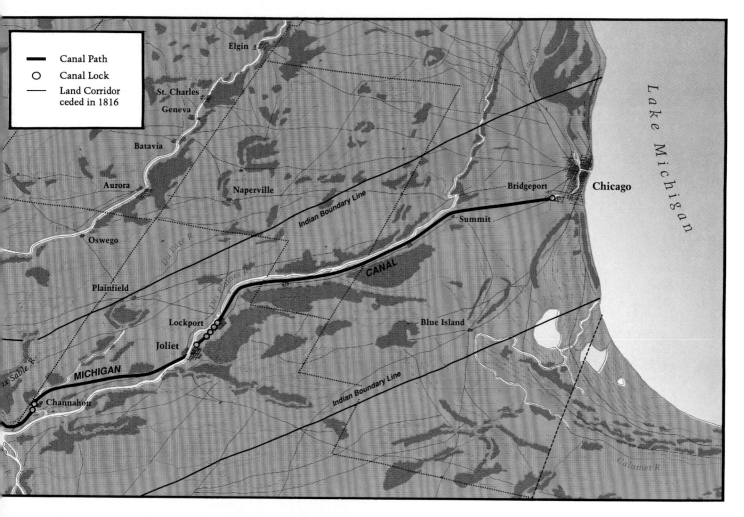

130. Aerial photograph showing the Desplaines River, the Sanitary and Ship Canal, and the Illinois and Michigan Canal near Lockport, 1954 (University Library, Champaign-Urbana)

Chicago to La Salle, about one hundred miles, in no more than twenty-four hours. The packets were, moreover, much more comfortable than the contemporary stagecoaches, with their jolting passage over the rough roads.

Figure 130 shows the Illinois and Michigan Canal just north of Lockport, where it curves sharply to the east. In the lower center are the tanks of a petroleum company. Bounding them to the west is the line of the Illinois and Michigan Canal; it looks thin, and the engineers seem to have been unable to make gentle curves. The abrupt changes of direction contrast with the smooth curves of the much broader and later Sanitary and Ship Canal, immediately to the west. Further to the west is the irregular bed of the Des Plaines River, which at some points is wide and marshy and at others narrow. On the ground it is often difficult to sort out these watercourses, but from the air they are quite distinct.

THE EMERGENCE OF CHICAGO · 93

The Western Terminus of the Canal at La Salle

FIGURE 131 SHOWS THE WESTERN END of the canal. The photograph was taken looking east; the Illinois River is in the foreground, the turning basin of the canal in the center. Further east are two sets of locks, and the canal winds away towards the top center. In 1865 the area of the turning basin was much less built up (figure 132), but at the beginning of the season it was full of barges. Now it is of course empty and indeed parts have become a marsh. The last lock before the turning basin has been reconstructed and can be seen in figure 131, just below the prominent bridge.

There were formerly many grain elevators along the western half of the canal, like those shown in figure 133, which were once to be seen at Ottawa. Their role was vital in the transport of grain, receiving it from farm wagons and storing it before it could be loaded onto the barges. One of these elevators survives at Seneca, and its inner workings appear in figure 134. Farmers unloaded their grain or corn at the ramp (1); the grain fell through the chute (2) into a cellar and was conveyed by scoops on a continuous belt up to the summit of the building (3). From there it was distributed by gravity into various hoppers (4), ready for further distribution into barges (5) on the canal. The Seneca elevator, with its massive wooden construction, is an impressive example of the industrial architecture that the emergent grain trade produced.

Looking at this building, we are reminded of the role the canal played in opening up the farmland to each side of it. This kind of elevator was built at frequent intervals along the canal, so that farmers within reasonable range could be sure of a place to store their corn or wheat, before shipping it to the vast and ever-growing Chicago market. Agricultural products also came north from centers like Peoria, and this trade formed part of a much larger exchange system, in which the agricultural products of the hinterland were bartered for commodities like timber and eastern hardwares, which came through the lakes and into the canal. The importance of the canal in initiating and maintaining this trade is shown by the number of towns that sprang up alongside it and that in many cases survive today.

131. Aerial photograph of the western terminus at La Salle, October 1988

132. Photograph of the La Salle
turning basin in 1865 (Illinois State
Historical Library)

APRIL 9-1865

HOSSACK

133. Elevators
along the canal in
Ottawa, about 1865
(Illinois State
Historical Library)

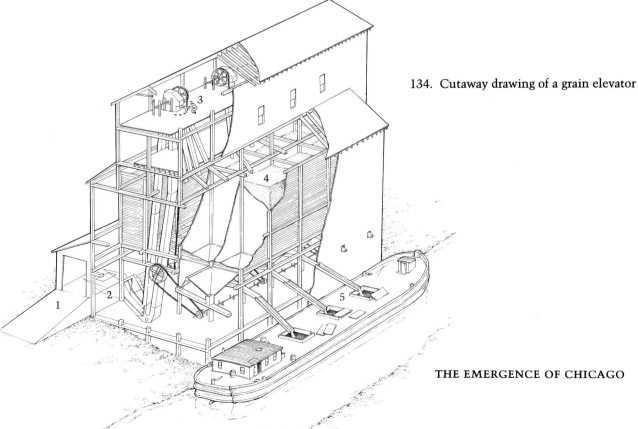

134. Cutaway drawing of a grain elevator

135. Aerial photograph of the lock and aqueduct at Aux Sable, April 1987

Canal Operations at Aux Sable and at Channahon

THE WELL-PRESERVED LOCK AT AUX SABLE, just west of the junction between the Kankakee River and the Illinois River, appears in figure 135; the lock's name goes back to the days of the French voyageurs. The outline of the towpath can be seen in the aerial photograph on the left-hand side of the canal. The reconstruction in figure 136 has identified the towpath, has taken out the later fish weir, and has restored the bottom set of gates.

We see a barge going downstream, towards La Salle. The boat was floated into the lock with the upper gates open; now they have been closed. The water will be let out of the lock so that the lower gates can be opened, and the barge will float out into the canal. The mule team waiting on the bank will take up the load again once the barge is through.

One difficulty of this system was the single towpath, upon which tows going in opposite directions often met. There ensued a carefully orchestrated maneuver, in which one of the teams moved to the outside of the path and dropped its towlines on the path and to the bottom of the canal, so that the other

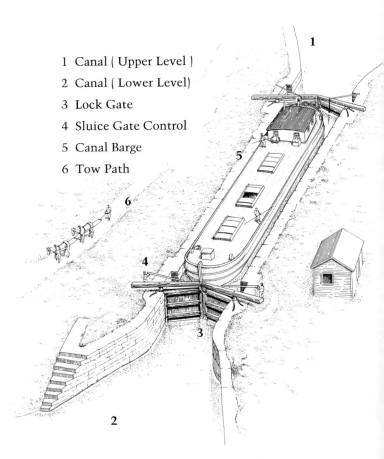

1 Canal (Upper Level)

2 Canal (Lower Level)

3 Lock Gate

4 Sluice Gate Control

5 Canal Barge

6 Tow Path

136. Diagram of the lock at Aux Sable

team could proceed over and past the first team's lines. This delicate operation was also carried out when a fast barge—a passenger packet, for instance—had to pass a slow one.

Note in the upper right corner of figure 135 the place where the canal (and the parallel road) crosses the Aux Sable Creek. The canal is carried over on one of the system's two aqueducts, which must have been a novel sensation for the barge's passengers. The Aux Sable Creek was quite narrow and easily bridged, but the engineers found a more difficult problem when they had to cross the much wider Du Page River at Channahon. Their ingenious solution is shown in figures 137 and 138. A barge coming down from the north (top) would lock down into the Du Page River by the lockkeeper's house (white in the photograph). Then it would be pulled across the river by its team walking along the top of the weir, before locking up into the canal at the southern end. This cheap and ingenious solution to a difficult problem may still be savored on the site, which is one of the best preserved on the canal; it is also one that can best be understood from the air.

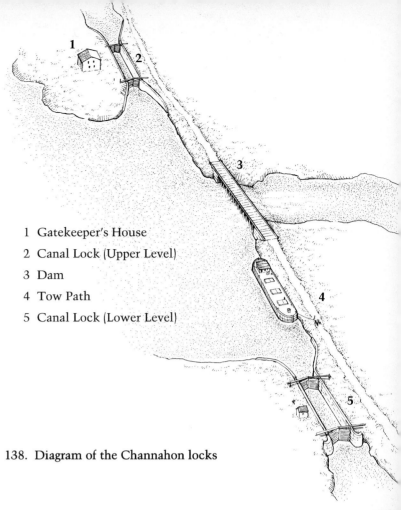

1 Gatekeeper's House
2 Canal Lock (Upper Level)
3 Dam
4 Tow Path
5 Canal Lock (Lower Level)

138. Diagram of the Channahon locks

137. Aerial photograph of the locks at Channahon, October 1985

139. Aerial photograph of the Bridgeport area in 1938 (University Library, Champaign-Urbana)

140. Detail from USGS 7.5-minute series, Englewood quadrangle (photo revision, 1972)

The Eastern Terminus at Bridgeport

THE AREA WHERE the diagonally aligned Illinois and Michigan Canal intrudes into the checkerboard of the Chicago street pattern became a huge center for commodity exchanges. Barges, often carrying grain, came east on the canal (figure 141) and along the river (top right) to the grain elevators. There they discharged their cargo, which might then leave Chicago by lake schooner, and returned to the numerous wharves found north of the river at the point where the canal joined it (figures 139–41).

These wharves carried a wide variety of goods, often brought in from the lakes and from the East by schooner. They were particularly important for their lumber, which southbound barges often loaded for their return journey. Our drawing shows the area about 1858, when it was alive with schooners and barges. Eventually, as the aerial photograph shows, the tongue of land between the canal and the river became a huge railyard, as the power of the railroads was added into the whole entrepot system. In the end the little canal was completely effaced and is now buried beneath I-55.

The line of the canal and of Archer Avenue, which ran alongside it, shapes the pattern of streets in a large part of Bridgeport. Note in the aerial photograph and on the modern map that only a few of the wharves remain, but the pattern of streets aligned with the canal has survived. This was the hub of the great exchange engine that first established the prosperity of Chicago.

141. Reconstructed view of the Illinois and Michigan Canal and Chicago River about 1858

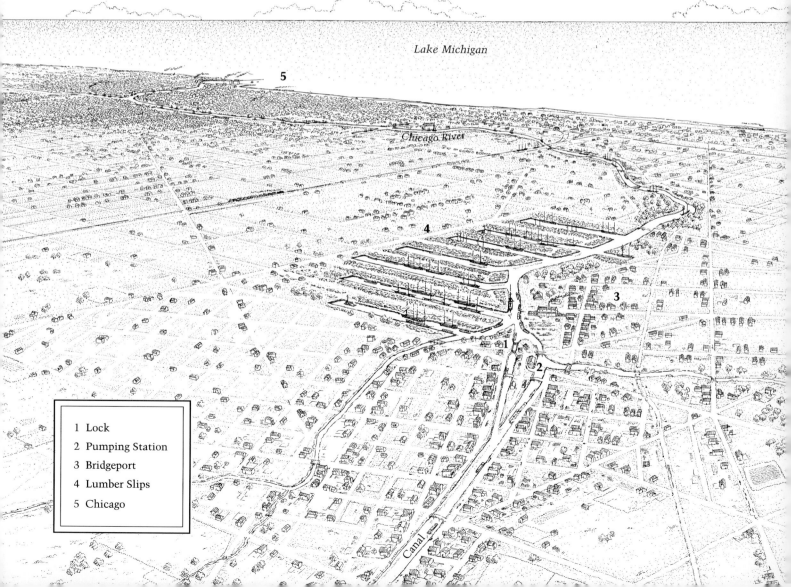

Lake Michigan

Chicago River

1 Lock
2 Pumping Station
3 Bridgeport
4 Lumber Slips
5 Chicago

Canal

A Bird's-eye View of Chicago in the 1850s

IGURE 142 SHOWS CHICAGO at the height of its development as a port and canal town, before the railroads had made any serious impact. Here we see the city on exactly the other side from figure 126. We are looking up the mouth of the river, which is lined with factories and grain elevators. A side-wheeler is approaching the harbor entrance from the northeast (note the new lighthouse on the tip of the northern arm), and to the south are two small schooners under sail, a large schooner with sails furled, and a steamboat with an auxiliary sail.

As the boats enter the river they come to the drawbridges, which are raised to let them pass. Some vessels proceed right down the South Branch (to the left) to Bridgeport, here out of sight, while others take the North Branch. At the mouth of the river on the south bank, space has been cleared for two huge grain elevators, which will long be conspicuous landmarks—the Sears Towers of their day.

Chicago is shown here at a time of transition. Fort Dearborn and the old lighthouse were still visible but would soon be gone, as would the open prairie, buried under the balloon frames. The city was beginning to influence national life: the flow of produce from the rich bottomlands of the Midwest, which hitherto had gone without question to the great metropolis at New Orleans, began to be diverted northwards to the new emporium on Lake Michigan. This process began before the advent of the railroads, and once they came to Chicago—often with a little political persuasion—the whole development gained a momentum that became irresistible, so that all Chicago's urban rivals were left far behind.

142. David W. Moody, View of Chicago about 1853 (after the drawing by George J. Robertson; Chicago Historical Society)

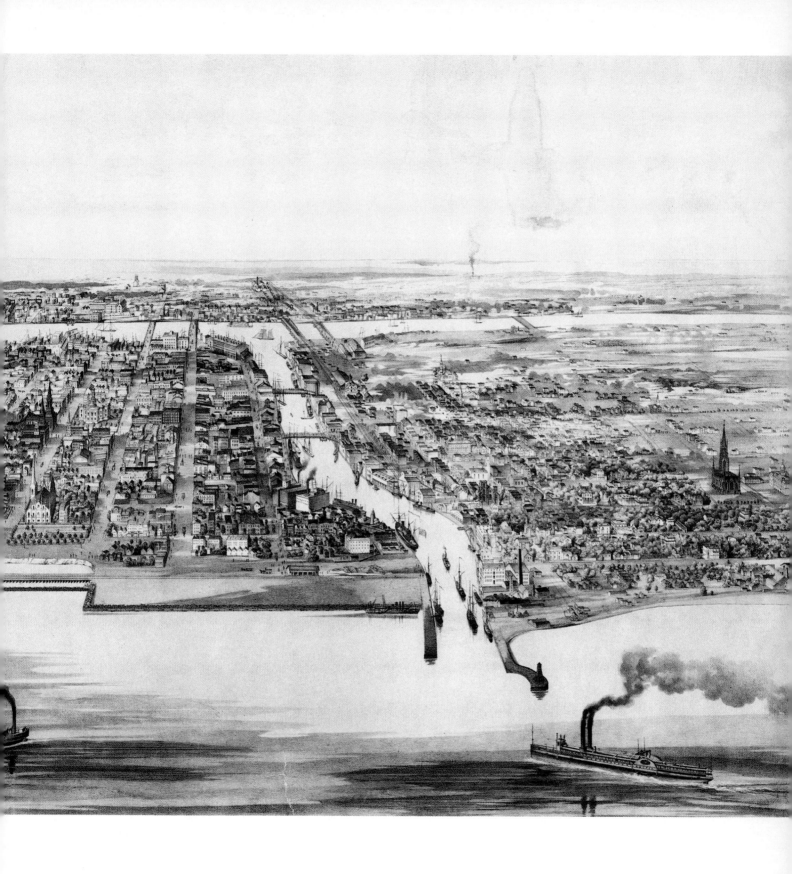

7

CHICAGO AND THE EXPANSION OF THE NORTHEAST

143. Detail from the bird's-eye view of Geneva in 1869, lithographed in Chicago (Geneva Public Library)

Introduction: The Metropolis and the Hinterland

THE LAND BOOM that accompanied the building of the Illinois and Michigan Canal was not confined to Chicago, for many of the immigrants arriving in its port continued into the northeastern part of Illinois. Newcomers from Europe and Canada joined an increasing flow of people from New England, New York, and Pennsylvania.

The most travelled routes, to judge by the distribution of inns, were to the northwest, leading to farmland around Elgin and Rockford and onward to the lead mines of Galena. Due west some inns were on roads leading to the Naperville settlement and the farmland of Du Page County. Another string of inns followed roads leading southwest along the canal and the Kankakee River and onward to the growing region of towns and farms along the canal corridor. From there the roads continued on to the emerging corn belt around Peoria and Springfield.

Much of the land into which the settlers were moving was prairie, but the introduction of John Deere's new steel plow in 1837 gave the individual farmer the means to break up the prairie sod, and inventions like McCormick's mechanical reaper let production go forward on a scale that supported exports rather than mere subsistence. Corn and grain were stored in elevators at many convenient points, especially in the huge Chicago elevators, which fed the export trade. There was also some local milling; in this period before steampower was widespread, every rivulet was examined to see if it would take a water mill. Du Page County even had four windmills, and another one was at Peotone.

Up to 1848 the road network and the Illinois and Michigan Canal had to carry all this increased activity, but in that year the Galena and Chicago Union Railroad began its first runs from Chicago west to Naperville. This new mode of transport made possible the settlement of the highly productive but somewhat isolated farmlands of the central prairie, doubling and then redoubling the production and export power of Illinois agriculture and simultaneously creating huge new markets for manufactures.

This process of expansion led to the creation of many new towns, often at mill sites. Geneva, shown roughly from the south in figure 143, was one such town. Notice the road bridge just to the north of the island in the Fox River, and beyond that the milldam, which provided water for the mill on the west bank. The town was laid out not quite north-south, but paralleled the river.

All this settlement and economic growth led to the emergence of a remarkably prosperous region, which has retained much of its economic buoyancy to the present day. But it also led to the destruction of much of the native vegetation and to continuing deforestation. Many of the beautiful groves with their abundant wildlife, so remarked upon by early visitors, have been destroyed, like most of the unique prairie.

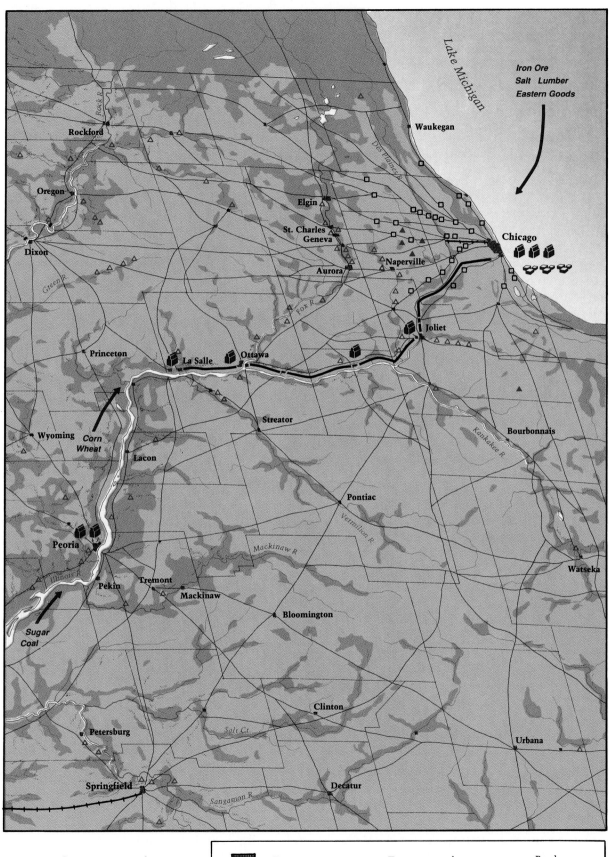

Lake Michigan

Iron Ore
Salt Lumber
Eastern Goods

Rockford

Waukegan

Oregon

Elgin

St. Charles
Geneva

Dixon

Aurora

Naperville

Chicago

Joliet

Princeton

La Salle   Ottawa

Streator

Bourbonnais

Wyoming

Corn
Wheat

Lacon

Pontiac

Peoria

Watseka

Pekin   Tremont

Mackinaw

Bloomington

Sugar
Coal

Clinton

Urbana

Petersburg

Springfield

Decatur

144. Aerial perspective of
northeast Illinois about 1850

	Town		Taverns around Chicago		Road
Grain Port			Water Mill		Canal
Lumber Port			Wind Mill		Railway

Immigrants from the East and Europe

Just as Morris Birkbeck, disillusioned with the state of England after the Napoleonic wars, had come to southern Illinois in the 1810s, so throughout the century a steady stream of other Europeans poured into the newly opened parts of North America. Often they were simply enterprising people in search of a new life, but sometimes they were fleeing difficult circumstances back in Europe. So it was with many of the continental Europeans who came to Illinois in the 1830s and 1840s. In France, Germany, and Italy these were turbulent years politically, culminating in the general revolts of 1848; on the whole the existing regimes held their own, so that immigrants tended to have rather "advanced" and even democratic views.

The British Isles were not subject to the same degree of political turbulence at this time, but the population was expanding rapidly as was the economy, and so younger sons often went to seek their fortune away from the narrowing confines of Great Britain. Ireland was a special case; there the Catholic population had never ceased to chafe under English rule and to seek refuge in emigration. Many Irish families had come over when the canals provided work, and in 1845 the island was stricken by a terrible po-

tato blight that drove as many as one-fifth of the people overseas, most of them to the United States.

Most immigrants to Illinois at this period came from North America. Canada had much political turmoil in the 1830s, as the Anglican establishment vainly tried to maintain its hold over economic and political life, and some Canadians who emigrated came to Illinois. So did thousands from the eastern seaboard of the United States, some shaken by the

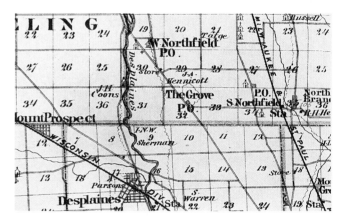

146. Detail from the *Atlas of the State of Illinois* (Chicago, 1876; Newberry Library)

145. Aerial photograph of the Grove, Glenview, June 1985

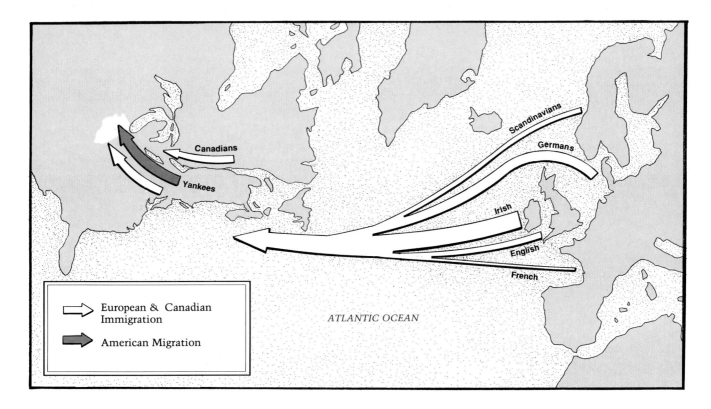

147. Map showing immigration to Illinois, 1830–50

financial crash of 1837, some leaving impoverished soil, and some simply seeking adventure in the West.

Many of these immigrants from the East were highly educated professionals like John Kennicott, who arrived in Cook County from New York State in 1836. He had qualified as a doctor at Fairfield and was a skilled horticulturist. He chose a delightful house site, on rising ground a little to the east of the Des Plaines River, and the house he built in 1856 is there today (figure 145). The land to the north is now a forest preserve, so that a well-chosen aerial view still gives us some idea of what it was like to come and settle on the edge of an Illinois grove in the 1830s. In his early days at the Grove, as he called his house, Kennicott would be visited by Indians, who camped on the lawn that is in front of the house.

In the 1830s and 1840s the forest behind the house and the prairie in front of it were full of animals and birds and the original plants; Kennicott became an expert on these and also pursued his horticultural interests, which led him to prominence in the Illinois State Agricultural Society. He was a leader in the movement to found a state university; he died in 1863, and his sons Charles and Robert carried on his work.

Charles greatly expanded his father's horticultural business; Robert became a noted scientist and was a founding member of the Chicago Academy of Sciences. He worked for the Smithsonian Institution

148. "Emigration to the West" (Robert Sears, *A New and Popular Pictorial Description of the United States*, New York, 1848; Newberry Library)

on such projects as recording the flora and fauna of pre-European-settlement Illinois; he travelled with the crew building the Illinois Central Railroad, recording the wildlife in the area around the track. He died in what would soon be called Alaska, on a scientific expedition, when he was only thirty. Even though the Grove is now surrounded by suburban sprawl, a visitor can still imagine Dr. Kennicott's pioneer family, his boys being formed, as they grew up, by the wealth of wildlife that surrounded them.

Settlement Patterns in Kane County

KANE COUNTY LIES IMMEDIATELY TO THE WEST of Chicago, with the Fox River on its eastern boundary; it offers an interesting variety of settlement patterns of the 1830s and 1840s. Figure 149 is based on an 1860 landownership map of Kane County, which we have colored to show the original forests and prairies; major settlement patterns are also indicated. Figure 150 is a satellite image of the same area.

The earliest groups to settle Kane County were the Irish, Welsh, and Scots, labelled in figure 149. Many were probably canal workers and their families; their concentrations in this figure correspond closely to the areas with wooded watercourses and to the irregular pre–Government Land Office survey field patterns in the satellite image.

The Irish settlers sought out the wooded areas in Rutland Township, in the north of the county; in the satellite image this area consists of broken farmland with small irregular fields. Perhaps some of the Irishmen had worked on the Galena and Chicago Union Railroad, which cuts right through their township. The areas of Scottish settlement were in Elgin and Plato townships; these were wooded river valleys in which the Scots established diagonal long-lots. These two patterns of tenure, so sharply at variance with the predominant rectangles of most of the county, were no doubt established before the arrival of the General Land Office surveyors, in the 1840s.

The Welsh settled Big Rock Township, and the long-lot pattern they established is most distinctive along the Big Rock River in figure 149; some of this pattern is still visible in the satellite image. The Canadian settlement of Kane County began in the early 1840s, and here we see an important shift in the settlement pattern. Canadians in Saint Charles Township maintained the traditional pattern of settlement along wooded watercourses, but in Burlington and Campton townships their farms have crept out onto prairie lands, and in the satellite image these areas are much more geometric. By the time of their arrival in the early 1840s the township-and-range surveys were well under way, and new technologies were making prairie farming possible.

The last and greatest wave of settlement to reach Kane County came from the East Coast of the United States in the 1840s and 1850s; by then large mechanized farms had become the norm, and they largely covered the rest of Kane County, almost exclusively occupying such townships as Burlington, Virgil, and Kaneville. These large rectangular holdings show up well in the satellite image.

We should remember that the information that we have used concerns land-*ownership* and not necessarily land occupancy (English financiers, for instance, were very active in buying land in the United States in the middle of the nineteenth century). Still, the remarkable correspondence between the satellite image and the national preferences of the map make us fairly sure that we have identified some of the original tenurial patterns that were deeply imprinted into the soil of Kane County.

149. Map of Kane County, showing settlement patterns, 1820–60

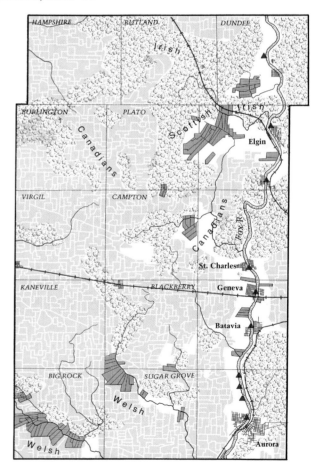

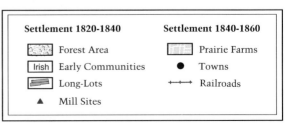

Settlement 1820-1840
- Forest Area
- Irish — Early Communities
- Long-Lots
- ▲ Mill Sites

Settlement 1840-1860
- Prairie Farms
- ● Towns
- ←→ Railroads

Most of the material in this section has been developed using an excellent study of structures and settlement patterns, which has been conducted by the Kane County Development Department. We are grateful to them for permission to use their material.

150. LANDSAT image of Kane County, September–October 1982 (Illinois State
Geological Survey and Northern Illinois University)

151. Aerial photograph of Garfield Farm, April 1988

Garfield Farm and the Agricultural Revolution

IN 1841 TIMOTHY AND HARRIET FROST GARFIELD brought their eight children from Vermont and settled at a promising spot about five miles west of Saint Charles, in Campton Township. There at the junction of two roads, they settled into a log cabin and began not only farming the 440-acre claim, but also taking in travellers. Both businesses prospered, so that in 1846 Timothy was able to construct the handsome brick house known as Garfield Farm, today a museum and working historical farm.

As Figure 151 shows, it is an elegant structure lying on one of the section roads laid out by the surveyors; large square fields surround it, and at the back are specialized outbuildings: a hay barn (1842) parallel to the house, a granary (1890) on the right, and a large dairy barn (1906) on the left. Garfield Farm worked about four hundred acres of mixed agriculture and almost from the start would have used machines like the Deere plow and the McCormick reaper, to win heavy crops from the abundant prairie. This type of farming is a far cry from the hardscrabble farms of early southern Illinois and came to produce exports at least for Chicago and sometimes overseas. As the farms prospered so did their sup-

pliers, and northern Illinois agriculture offered a market to factories producing farm machinery in Chicago, Moline, and Rockford.

Figure 153 is an illustration from the Menard County atlas of 1874. County atlases containing maps like this exist for virtually every county in Illinois. These atlases have not been studied as intensively as is desirable, but the preliminary results of work by scholars like Michael Conzen suggest that the maps are reasonably accurate. Figure 153 shows

152. John Deere's "Improved Clipper" (*Country Gentleman*, August 1857; Newberry Library)

a farm somewhat larger than the Garfield Farm, but otherwise similar. Roughly half the area is in pasture, with the rest divided almost equally between woodland and arable fields, in which the crops are corn, wheat, and oats. In the middle cornfield may be seen two plow teams, and in the small central wheat field is a mechanical reaper. This whole substantial venture was directed from the farmhouse, seen with its garden and orchard at the left center, on the edge of the wood and by the little watercourse.

Menard County is in the extreme southwest of the area discussed in this chapter, but the scene shown here was typical of the central prairie areas, with their large square fields lying within a rectangular road system. This is the type of farm particularly associated with the immigrants from the eastern states, who used their capital to establish these sophisticated agricultural units on the most productive land. The entire view shown in figure 153 represents only two or three of the large squares shown on the satellite image, figure 150.

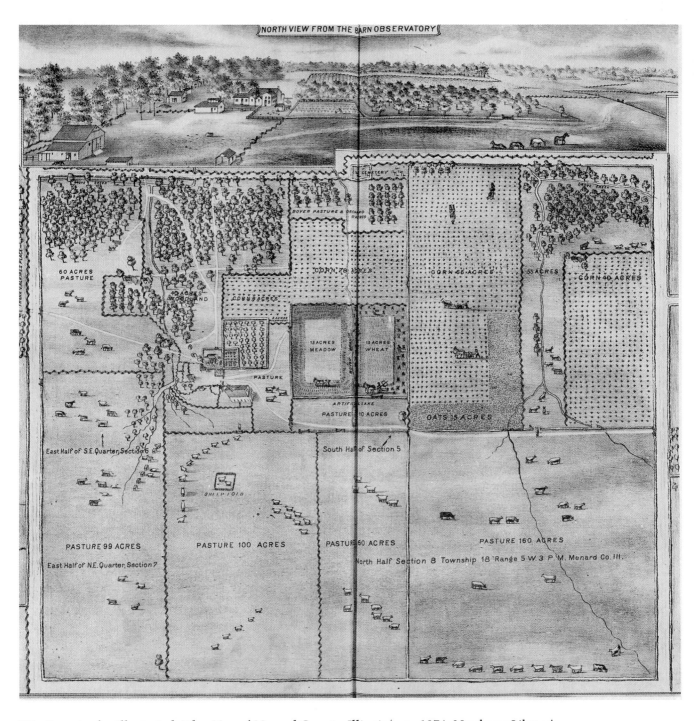

153. Farm in the *Illustrated Atlas Map of Menard County, Illinois* (n.p., 1874; Newberry Library)

154. Aerial photograph of the Graue Mill, April 1986

The Graue Mill and Other Water Mills

WHEN THE ILLINOIS INDIANS wanted to grind corn, they used a hand-operated pestle. During the French occupation, gristmills operated by animals were common, and after about 1830 steam power began to be used for milling. From about 1790 onwards, however, with the slow penetration of the state by easterners, another type of power was also used: that of the streams and rivers. The General Land Office surveyors were instructed to make special note of sites suitable for water mills, and by the 1830s there were almost fifty of them in Illinois. Others continued to be built for a decade or two, but it seems that very few survive today in anything like their original form.

One of these we have already mentioned (figure 73) in discussing New Salem, where the Carpenter Mill has been restored. Two others are the Baltic Mill on the Kishwaukee outside Rockford, and the Graue Mill on Salt Creek at Hinsdale. Water mills may be classified by the type of wheel they use, either undershot (the water driving the lowest of a set of paddles), overshot (the water driving the uppermost of a set of paddles), or turbine (the water driving propeller-like vanes under pressure). Broadly speaking, undershot wheels are the earliest, followed by overshot, and then by the more sophisticated turbine.

All mills need a good water source, with a certain flow. Often they were located on rivers at sites that were also suitable for fords, and so became the focus

155. Ground view of the Baltic Mill, Belvidere, June 1988

of settlements. The Graue Mill used a relatively minor creek, penning its waters up in a millpond (top of figure 154). From there the water ran down a sluiceway until it hit the undershot wheel, just visible on the right-hand end of the building. As the wheel revolved, it turned the huge millstones, which ground the wheat or corn into flour. The Graue Mill was built by Fred Graue from Hanover, Germany, between 1847 and 1852 and continued to function until the end of the century; it is one of the few well-authenticated Illinois stops on the "Underground Railway," by which former slaves from the southern states were spirited north to freedom in Canada.

The Graue Mill has long been preserved and is open to visitors. The Baltic Mill, on the other hand, is only now in process of restoration. Built about 1845, it works with a turbine, invisible from the exterior. Water mills of all types were indispensable in the early phase of settlement, for grinding wheat (corn can be used without processing it in this way) and for sawing timber. But the sawmills tended to be less permanent; they moved in the vanguard of settlement and, when an area's timber was exhausted (and often, its streams choked with sawdust), could be dismantled and moved on.

Milling breaks up cereal grain into meal and removes unwanted matter. After threshing and winnowing in the barn, farmers would bring sacks of grain kernels (figure 156) to the mill. There the outer coating of bran (1) was separated from the endosperm (2) and germ (3), which were ground into flour.

The old process mill was driven by a waterwheel (4), which transmitted its power to a gear arrangement (5) that revolved the millstones (6) and worked such equipment as pulleys (7) and bolting machines (13, 14). Grain delivered to the mill was poured into a rolling wire screen (8) that removed any large particles. From there it fell in front of a Dutch fan (9) that blew away the lighter dust. The cleaned grain was raised to hoppers (10) that fed the grindstones (6). There the outer bran was torn off and the heart of the kernel ground into flour. The resulting meal was collected below (11), lifted to the upper floors, and shovelled out into long bins (12) to cool and dry. Finally the meal was poured into the bolting ma-

chines. These were spinning cylinders covered with cloth (13) or wire mesh (14) that separated the flour from the bran. The final products were bagged or barrelled at the end of long spouts (15).

Figure 156 also shows the grinding room with a millstone case in the foreground (16). In the background a case has been removed and the upper millstone (17) raised and flipped by a crane (18) in order to be sharpened. Millstones in constant use were sharpened once or twice a week with iron picks (19). The stone used to make grindstones had to be a particular type, and the best ones came from France.

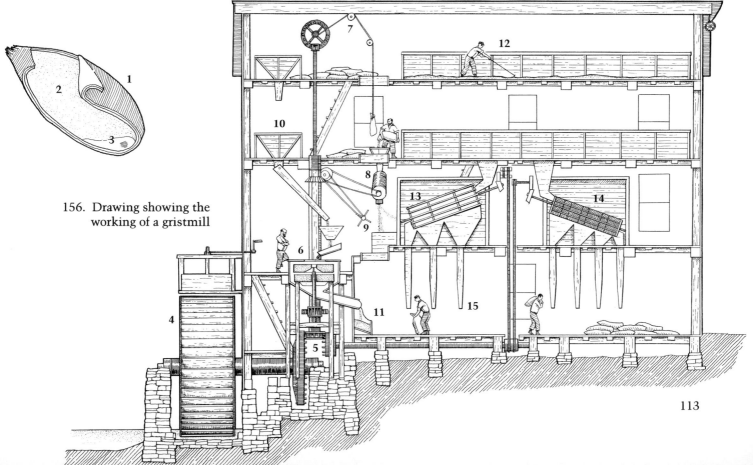

156. Drawing showing the working of a gristmill

157. Aerial photograph of the Fabyan Mill at Geneva, October 1988

The Windmills of Du Page County

WHEN THE EUROPEANS CAME TO NORTH AMERICA, they brought with them the technology of windmills and erected a good many in the East as early as the seventeenth century. The four that were constructed in Du Page County (see figure 144) were built about the middle of the nineteenth century by Germans, though they are known as "Dutch mills." The Holstein Mill in Bloomingdale and the Heidemann in Addison have disappeared, the first destroyed by a tornado in 1899 and the second by a fire in 1958. But the Fischer Mill and the Old Holland Mill have miraculously survived.

The Fischer Mill is on its original site by Grand Avenue, one mile east of York Road; it is plainly visible from I-294 to the west, about three miles south of O'Hare Airport. The mill was built between 1847 and 1850 by Frederick Fischer, following the style that was current in Europe. Figure 158 shows it as it looked in the early 1920s, when it was still capable of producing about thirty barrels of flour a day. It has a stone base and a wooden superstructure, surmounted by a cap that holds the huge sails and can swivel to catch the wind. This remarkable structure survived because in 1925 the land on which it stands was bought by the Mount Emblem Cemetery Association, which has since that time looked after it.

The Old Holland Mill, built by Louis Backhaus in the 1850s, was of a similar size and type. It worked at its site on the York Road for about half a century but by 1914 was falling into disrepair. By great good fortune, its plight came to the attention of "Colonel" George Fabyan, a wealthy businessman originally from Boston, who had a superb estate on the Fox River at Saint Charles. Fabyan was an eccentric in the grand style; among other enthusiasms he was keen on producing naturally milled flour. He had the Old Holland Mill taken down, transported to a site on his estate, and reerected. This was done with the

158. Photograph of the Fischer Mill, ca. 1921 (Percy Sloan Collection; Newberry Library)

greatest care, down to making sure that, where possible, woodgrains on the floorboards matched.

Today the mill stands in a clear space above the river (figure 157). It was for a time used as living quarters and in 1915 housed William Friedman, a graduate student in genetics from Cornell. Fabyan at first had Friedman conduct experiments like planting oats at different phases of the moon, to see if that made any difference to growth. But Friedman soon got involved in a different kind of investigation. Together with his wife, Elizabeth Smith, he worked for Fabyan on projects in cryptography, to such effect that they became leading experts for the federal government in both world wars. Fabyan's Riverbank Laboratories conducted many other kinds of investigations, but none was as successful as its cryptographic activities, begun by Friedman in the windmill.

Roads and Taverns around Chicago

THE CHICAGO PORTAGE WAS A FOCUS of trails in the time of the Indians. Many of these trails, which tended to follow the higher ground and so to be relatively dry, were adopted by the early European settlers. Still, it was one thing to follow a trail on foot or on horseback, like the Indians, and another to drive a heavy wagon along it. East-west communications were particularly hampered by the pattern of the rivers, which all run north-south, from the Chicago River in the east to the Fox River in the west (figure 160).

Making roads to serve the growing number of farms was slow. The best roads were made by laying a substantial foundation and topping it with gravel, but this could be an expensive business, particularly in the many boggy patches, which seemed simply to absorb whatever was laid down. Local farmers became skillful at keeping the roads and bridges open, but even their best work was often not equal to the conditions brought on by the spring rains, which turned the roads into seas of black mud, littered with broken-down and abandoned vehicles.

In the early 1850s, when this map was produced, an attempt to overcome the problems was made by building plank roads, a technique that seems to have come from Russia via Canada and that involved laying three-inch planks crosswise on wooden stringers bedded in the ground. Companies were formed to build these roads, which were sixteen feet wide and cost about $2,000 a mile; tollhouses were at five-mile intervals, and a few have survived. Four of these roads radiated out from Chicago, and three of them may be seen on our map. Their vogue did not last long, for the wood rapidly decayed in the marshy ground, and at the best of times they were extremely rough and noisy. So the settlers had to go back to the expensive but more permanent way of laying a heavy foundation.

159. Aerial photograph of Stacy's Tavern, Glen Ellyn, May 1989

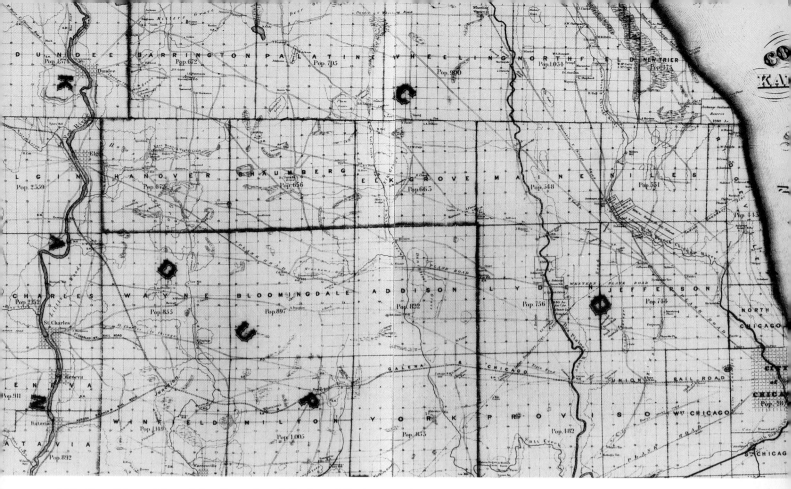

160. Detail from *Map of the Counties of Cook and DuPage* ... (James Rees, Chicago, 1851; Newberry Library)

A wide variety of traffic used the roads. Conestoga wagons carried immigrants hundreds of miles from the East and served to transport the produce of their new farms. People like doctors and parsons were beginning to drive lightweight carriages. Farmers drove huge herds of animals to the Chicago market, for at this time animals were still transported in this time-honored way. Stagecoaches from the mid-1830s onwards had regular services that linked Chicago with the western towns. Until their demise with the coming of the railways in the 1850s, these stagecoaches were no doubt the most picturesque vehicles on the road, with their well-sprung chassis, elegantly painted bodies, and lively teams of horses.

The network of stagecoach routes encouraged the establishment of more taverns, where horses could be changed and travellers could refresh themselves. Often these taverns were by river crossings, but sometimes they grew up at the intersection of main routes. One such tavern was Stacy's (figure 159), which was built where a north-south road meets the fork in the road from Chicago, one branch going to Saint Charles and the other to Geneva. Moses Stacy had come from Massachusetts and in 1846 built his inn at what would become known as Stacy's Corners; it survives to the present day. People tended to settle near the taverns, which were centers for information as well as of social life; this often gave them a natural clientele that let them survive the coming of the railroads, which in general diverted custom from the older routes and halting places.

161. Aerial view of Stacy's Tavern in the 1840s

162. Aerial photograph of the schoolhouse now at Palos Hills, April 1986

Post Offices, Churches, and Schools

AMONG THE HEAVIEST USERS of the road network were the contractors hired to carry the federal mail, stopping at post offices like the one at South Grove shown in figure 163. It is almost impossible for us to imagine the slowness of the system in the 1830s, when from Chicago it took a letter more than sixty days to reach New York and over a week to arrive at Springfield. For outlying areas like Green Bay the mail went by foot. In that particular case an Indian guide accompanied the carrier, and the round trip took one month.

The post offices were distributed more or less logically along the major highways, with fresh agencies added as the pattern of settlement seemed to call for them. Churches and schools, on the other hand, were scattered among the rectangular countryside in the most unpredictable fashion. In figure 163, a map of part of De Kalb County, the churches are the little buildings marked with a cross and the schools are

similar symbols, with a bell tower. In New England, where there was a considerable degree of religious homogeneity, churches tended to be sited more or less at the center of communities. But in early Illinois there was an extraordinary profusion of sects, so that the churches were often small and widely scattered. They were also often very simple in their construction and impermanent, as they might easily be abandoned if their fragile demographic base changed for the worse.

Schools were equally numerous and scattered. It was not until the 1850s that the state government began to exercise serious supervision over the education system, organizing school districts which in theory (but almost never in practice) could sustain their activities from the proceeds of section 16, set aside in each township for educational purposes (see figure 63). The siting of the schools was the subject of much contention and calculation. It was good for a farmer, for instance, to have a school close by, so that his children were spared a long walk and could be back early to help on the farm. On the other hand, if the school were too close, the pupils might damage his crops in their boisterous play.

Out of innumerable calculations, then, came an apparently incoherent pattern of schoolhouses. The buildings were at first rough and often unsuitable, built by farmers more skilled at lodging animals than catering for the needs of learning. But as time went by a more or less standard building came to be adopted, like the one shown in figure 162. This is a typical Illinois schoolhouse of the later nineteenth century, with one large room, ample windows for light and ventilation, and a belfry in which hung the bell that called the scholars to their labors.

This particular "little red schoolhouse" was built in 1886 and was moved to its present site in Palos Hills in 1952. It is now a nature center, with many publications for sale and a considerable collection of indigenous mammals and reptiles. Many of the scattered schools were abandoned, as demographic patterns changed and it became possible to use buses to bring students from long distances. But many of the buildings still exist, often converted into residences or storehouses.

163. Detail of De Kalb County showing churches and schools in red. (*Atlas of the State of Illinois*, Chicago, 1876; Newberry Library)

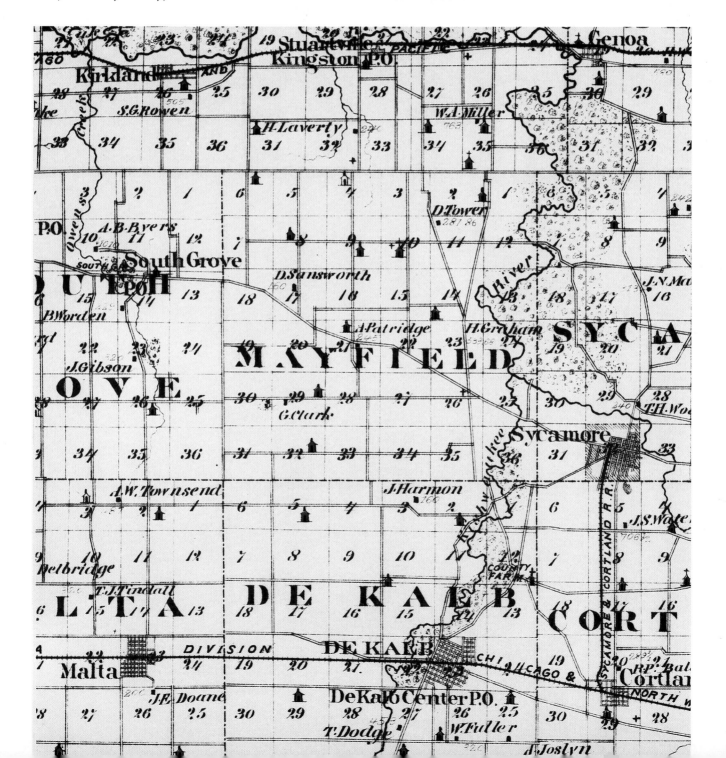

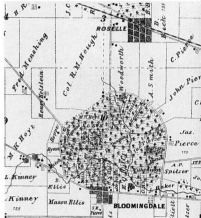

Settlement and the Dwindling Groves

IN THE 1850s, AS FIGURE 144 SHOWS, there were extensive areas of woodland in the northeast. By the 1980s much of this beautiful and ecologically desirable forest was destroyed. Figure 166 shows the pattern of this destruction in Kane County, which is in fact one of the less affected areas. Along the Fox River the dense groves have mostly given way to urban areas, with congregations of trees found only in parks. In the agricultural area to the west, the trees have been almost entirely cleared in order to form fields. All in all, it looks as if each forested section is about one-sixth of its former area.

The same is true of most of the groves that once dotted the plain between the Fox River and Lake Michigan. At Bloomingdale was a grove of particularly distinctive shape, which in the 1874 map (figure 165) was well defined by the roads that almost

surrounded it. By 1980 this grove, still defined by the roads, had dwindled to the size shown in figure 164, which is a detail from the U.S. Geological Survey map revised to that date. We do not have an aerial photograph of the late 1980s, but it would no doubt show still more encroachment by houses and parking lots.

166. Aerial perspectives showing vegetation in Kane County in 1830 and 1988

167. Detail from aerial photograph of northern suburbs of Chicago, 1970 (Illinois Department of Transportation)

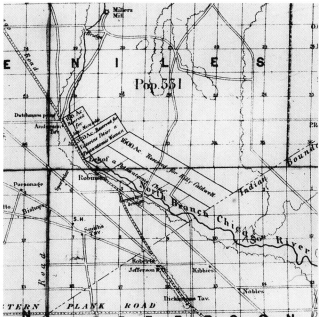

168. Detail from *Map of the Counties of Cook and DuPage* (James Rees, Chicago, 1851; Newberry Library)

The major survivals from the nineteenth century are to be found in the area's forest preserves. In the 1830s dense woodland ran along the North Branch of the Chicago River. As figure 168 shows, this area was in 1851 "Reserved for Billy Caldwell" and for Victorine Potier and Jane Miranda, whose smaller blocks lay to the northwest of Caldwell's large diagonal strip.

By the 1980s this Billy Caldwell Forest Preserve contained one of the few surviving large blocks of forest in the area of north Chicago. Figure 167 is a detail from an aerial photograph of 1970, showing the northwestern end of Caldwell's reservation. It is still possible to identify the three separate holdings. Caldwell's retains a large strip of woodland alongside the river, but the area to the north of this is covered by a subdivision. Potier's remains half in forest and the rest under houses; only Miranda's has survived more or less intact, in spite of some encroachment from the southwest.

Of course, it is not always parking lots and supermarkets that have taken the place of the groves. Sometimes they were replaced by well-treed suburbs, and there might even be a case for maintaining that, considering the great areas of prairie that are now under suburban housing, there are *more* trees in all in the area than there were in the 1830s. However, these trees are widely separated, so that they do not have the aesthetic effect of the groves, which the settlers found so attractive. Nor can they sustain the same patterns of wildlife, which included many species for which the shelter of the woods was essential.

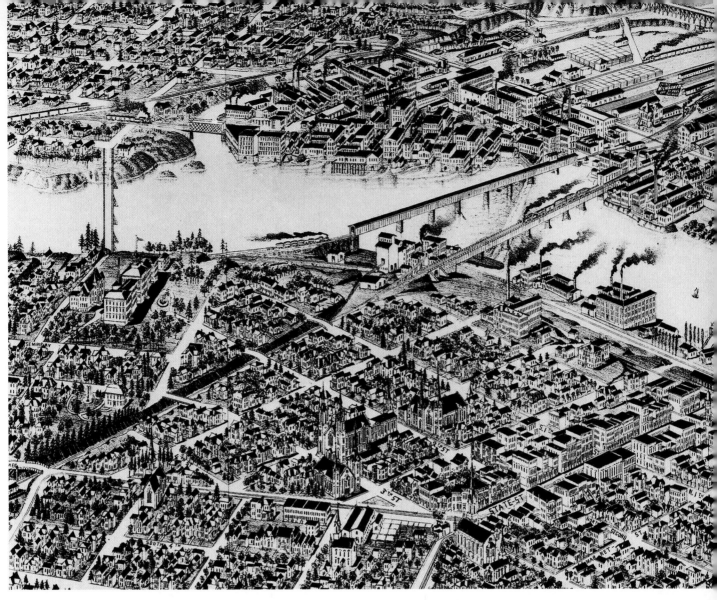

169. Detail from the *Perspective Map of the City of Rockford, Ill., 1891* (Newberry Library)

The Emergence of Rockford

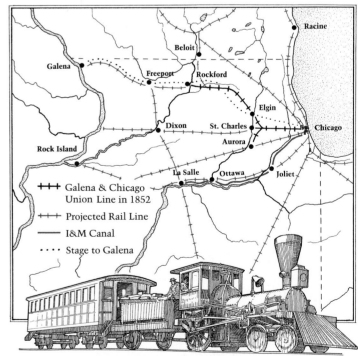

THE THRIVING NORTHWESTERN REGION, described in chapter 5, was linked to Chicago in the mid-1830s by a road used by teamsters hauling lead from Galena (the trip to Chicago could take six weeks) and by settlers moving west. In 1838 a stagecoach line began using this road. About midway, the route crossed the Rock River, and there the settlement of Rockford (first known as Midway) began to emerge. Figure 171 shows the site in 1839, when the coach line crossed the river at a ford, with a sawmill on a small tributary to the west and a small settlement, including a hotel and stage barn, on the east side.

The ford not only was the natural place to cross the river but also lent itself to the construction of a

170. (left) Map to show Chicago-Galena
stagecoach route and subsequent track of the
Galena and Chicago Union Railroad, 1852

dam, after which the powerful flow of the river could be channelled into large mills. The 1840s saw a dam built, and by 1859 the site looked as is shown in figure 171. Numerous mills were established along a great millrace (where Race Street would later emerge) provided power for various types of manufacture, including the furniture for which Rockford became famous. By now the town was substantial, with several hotels and a fair number of houses on both banks. Notice that the streets follow the line of the river in the center of town but revert to the north-south rectangular orientation in the outskirts.

A prominent novelty is the railroad line, which curves in from the north and crosses the river by the original ford, displacing the road bridge some way to the east. This line was the Galena and Chicago Union Railroad, which began building track westward from Chicago in 1848, reaching Rockford in 1851. This railroad was intended primarily to carry migrants westward, as part of the great system originating in the Atlantic ports, but in its early years it was chiefly instrumental in opening the rich farmland along its track, particularly in the Rock River valley, whose fertility Black Hawk had so praised.

Rockford was a great beneficiary of this expansion, as figure 169 suggests. This bird's-eye view shows the town at the peak of its expansion as a railroad and milling center, serving a large and rich agricultural area. In the milling district, turbine-powered mills (note the round apertures for the spent water) produced farm implements as well as furniture, while the railroad tracks branch off to serve lumberyards and grain elevators. Rockford had become a hub for the populated area of northern Illinois (figure 170), and beyond that it was part of a vast system, importing specialized necessities and luxuries from the East and from Europe and exporting eastwards the products of the region, some like grain in raw form, but others like furniture in a manufactured state. Rockford remains to this day a considerable manufacturing center and is the second-largest city in Illinois.

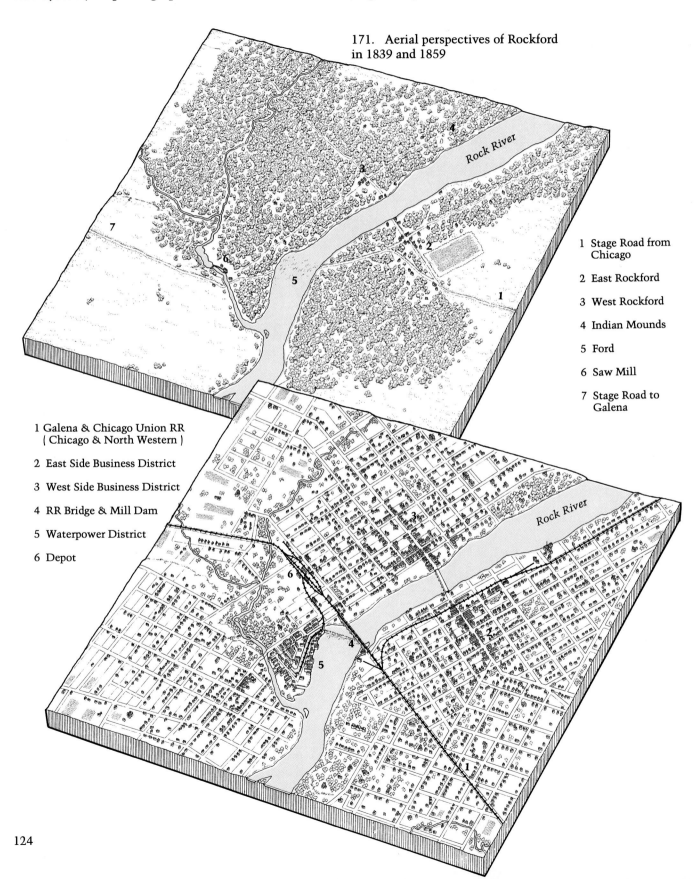

171. Aerial perspectives of Rockford in 1839 and 1859

1 Stage Road from Chicago

2 East Rockford

3 West Rockford

4 Indian Mounds

5 Ford

6 Saw Mill

7 Stage Road to Galena

1 Galena & Chicago Union RR (Chicago & North Western)

2 East Side Business District

3 West Side Business District

4 RR Bridge & Mill Dam

5 Waterpower District

6 Depot

8

RAILROADS IN ILLINOIS

172. George Parrish, modern painting depicting the construction of the Illinois Central Railroad (Illinois State Historical Library)

Introduction: Railroads from 1850 to 1880

FROM THE MID-1830S ONWARDS there were proposals to build railroads in Illinois, but it was not until the late 1840s that serious progress was made, with the Galena and Chicago Union Railroad (in black in figure 173). By 1860 the lines shown in red were in operation. The prominent north-south line belonged to the Illinois Central Railroad, which by then had the greatest length of track of any company in the world; the east-west lines belonged to other companies that would play a leading part in the economic development of Chicago and of the region. By 1860 there were nearly three thousand miles of track in Illinois, and the outline of the trunk routes was fixed; the next twenty years would see feeder lines extended into almost every township.

The concentration of railroad lines into Chicago was partly the result of intense political lobbying and partly the inevitable result of geography, for lines from the East had to pass by the southern end of Lake Michigan, where Chicago had the only satisfactory port, with all transshipment possibilities. A map of 1856 already has a sailing vessel off Chicago, with the legend "Direct by way of R. Saint Laurence to Europe"; although this particular dream had not yet come to fulfillment, the city's future as a world transport hub was already being fixed.

This development of railroad lines led to the opening up of vast areas of agricultural land in Illinois and to greatly increased economic activity in Chicago itself. It also had consequences on a national scale, for the fertile lands of the Midwest would no longer be tied in with the Mississippi Valley, as in the heyday of steamboats, but rather with the great markets of the East and Europe. Within one decade an east-west axis had replaced the north-south one, with political as well as economic consequences. When the Civil War came, in spite of the heavy mix of settlers from the South, Illinois naturally sided with the Union states to the east, its new economic partners, and formed a sort of spearhead, penetrating Confederate territory.

During the Civil War the southern terminal of the Illinois Central Railroad at Cairo became a huge advance base for Union forces, which were sustained by supplies brought quickly along the railroads from the inexhaustible and well-protected resources to the north. Chicago became a great arsenal during the war, providing not only foodstuffs to all kinds but also enormous supplies of uniforms and equipment. The city's economic growth was enormously stimulated, as was the development of the railroads that supplied its markets.

Once the war was over, the east-west relationship became even stronger, as Chicago became the starting point for ventures to tap the mineral and agricultural wealth of the West. Many of the railroads that had emerged in the Chicago region became leading contenders in this great westward movement, which eventually made Chicago the head of an economic empire stretching westwards as far as the Rocky Mountains.

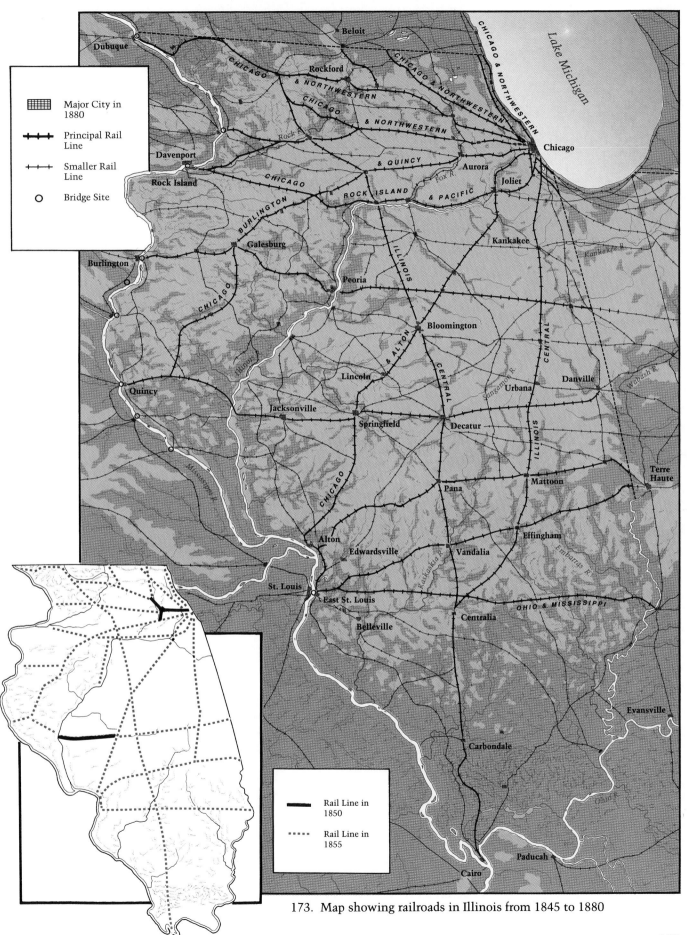

Legend

- Major City in 1880
- Principal Rail Line
- Smaller Rail Line
- ○ Bridge Site

Lake Michigan

Dubuque
Beloit
Rockford
CHICAGO & NORTHWESTERN
CHICAGO
& NORTHWESTERN
& NORTHWESTERN
CHICAGO & NORTHWESTERN
Chicago
Davenport
Rock R.
Rock Island
CHICAGO
& QUINCY
Aurora
Fox R.
Joliet
BURLINGTON
ROCK ISLAND
& PACIFIC
Galesburg
Kankakee
Kankakee R.
Burlington
CHICAGO
Peoria
ILLINOIS
Bloomington
Quincy
Lincoln
CENTRAL
& ALTON
Urbana
Danville
Wabash R.
Jacksonville
Springfield
Decatur
CENTRAL
Missouri R.
Illinois R.
Sangamon R.
Pana
Mattoon
Terre Haute
CHICAGO
Effingham
Embarras R.
Alton
Edwardsville
Vandalia
St. Louis
Kaskaskia R.
East St. Louis
Centralia
OHIO & MISSISSIPPI
Belleville
Evansville
Carbondale
Ohio R.
Rail Line in 1850
Rail Line in 1855
Paducah
Cairo

173. Map showing railroads in Illinois from 1845 to 1880

The Colonization Work of the Illinois Central Railroad

BY THE 1840s the only substantially empty part of Illinois was the central-eastern prairie region. It was the "colonization work," as Paul Gates put it, of the Illinois Central Railroad that during the 1850s filled this land with prosperous farmers, linked by rail with the Chicago market.

As early as 1837 the Illinois legislature had voted money for "an Act to establish and maintain a general system of internal improvement," the core of which would be a central north-south railroad. But a financial crash followed, and the impoverished state government lacked the kind of funds needed. In 1850 Congress granted public lands to the state to aid in the construction of a railway, rather as it had done in the case of the Illinois and Michigan Canal. But whereas the canal took only three hundred thousand acres, the railroad was allotted two and a half million acres, in a grant that was the first of its kind.

A Y-shaped system was planned, as shown in figure 174, beginning at Cairo in the south and branching at Centralia, with one line going to Chicago and the other to Galena. The even-numbered sections for six miles on either side of the right of way were assigned to pay for the work, which was carried forward at amazing speed (figure 172). By 1856 there were 705 miles of track, which had cost $25 million, more than the Erie Canal. Here and there, as the map shows, competing railroads crossed the Illinois Central tracks; the rule was that each train came to a dead halt, and then the first one there could proceed.

Using heavy and skillful advertising, the Illinois Central eventually sold land to about thirty-five thousand different families and so peopled a vast tract of central Illinois. Figure 175, looking towards Chicago, shows the right-hand branch near Paxton (many of the stations, like this one, were named after the engineer responsible for that section). The line seems to stretch off into infinity, with little towns at regular intervals to gather the tribute of the prairies into their huge elevators.

The Illinois Central was a pioneer in shipping exotic fruit as well as more conventional foodstuffs and in 1867, for instance, began sending strawberries from southern Illinois to Chicago under refrig-

eration. Eventually the Illinois Central, or the Illinois, Central, Gulf as it became in 1972, extended westwards across Iowa to Sioux City and southwards to New Orleans, but the core of its activities remained the Y-shaped lines in Illinois.

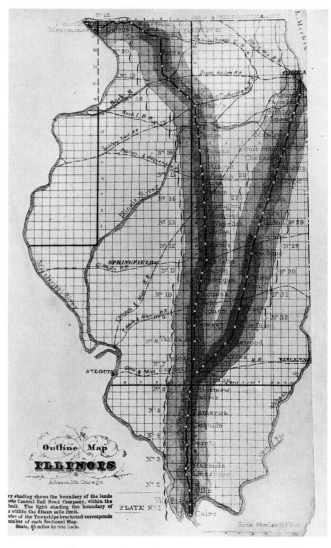

174. *Outline Map of Illinois* showing Illinois Central lands, 1855 (Newberry Library)

175. (opposite) Aerial photograph of Illinois Central track, September 1986

176. Aerial photograph of Cairo, September 1986

Towns along the Tracks

TOWNS AND RAILROADS WERE INSEPARABLE in the minds of nineteenth-century entrepreneurs, for each depended on the other. Thus Darius Holbrook, who bought land at Cairo in 1835, became the great booster of the Illinois Central. He thought that by linking Galena and Cairo, the latter would infallibly become a great metropolis (figure 176). It is situated at a promising place, where the Ohio and Mississippi meet, and in 1839 Holbrook succeeded in raising over a million dollars in London for Cairo's development. Alas, the junction of the two rivers is also a great place for floods, and very little of the money was recovered for the investors. One of these may have been Charles Dickens, who on a visit to the United States in 1842 wrote about Cairo as "a place without one single quality, in earth or air or water, to recommend it."

Of course, not all towns to which the railroad gave birth had such unfortunate histories. Many became thriving little places, often centered on a small rail siding that served one or two elevators, a stockyard, boardinghouses, and a lumberyard (figure 177). Sometimes the town attracted a population large enough to support churches and schools. The rail-road could destroy as well as create. A locality that seemed marked for success by a creek or wooded stretch would surely decline if the railroad bypassed it, and this was well understood by contemporaries, who waged fierce political warfare to attract railroads to their sites.

The shape of railroad towns is interesting. Occasionally their streets continued to follow the north-south and east-west orientation of the township system, and the railroad cut brutally into the existing pattern. More often the town was platted along the line of the railroad, as in figure 178, showing Nora in Jo Daviess County. Here the "Rail Road Addition" makes up the greater part of the town, though the "Original Town" is roughly aligned with it. Perhaps Nora was originally aligned with the Chicago and Galena Stage Road, which can be seen coming down from top left at the same orientation as the tracks of the Illinois Central. Nora in 1893 had a depot in the middle of town, with several warehouses and some stockyards alongside the tracks. There were many empty lots, often used as lumberyards, but also a hotel, a school, and two churches. It looked like a little railroad town that would survive, and indeed it has.

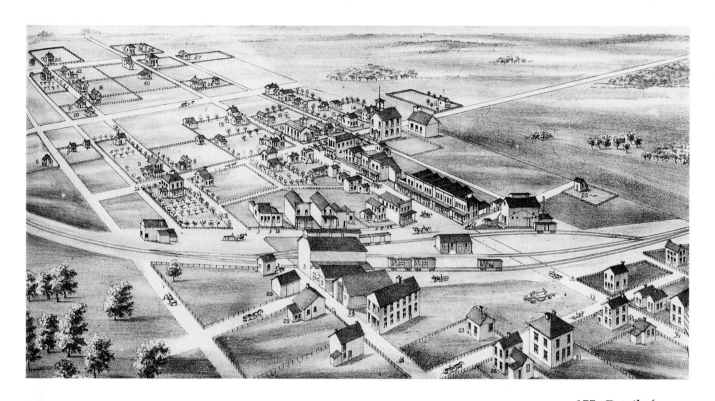

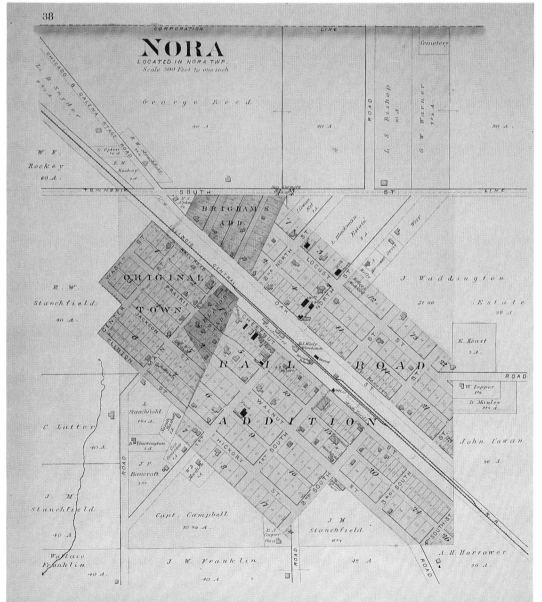

177. Detail of Elwood (*Atlas Map of Will County*, Chicago, 1873; Newberry Library)

178. Plan of Nora (*Atlas of Jo Daviess County*, Chicago, 1872; Newberry Library)

179. I. T. Palmatary, detail from a lithograph of Chicago, 1857 (Chicago Historical Society)

Chicago and the Railroads

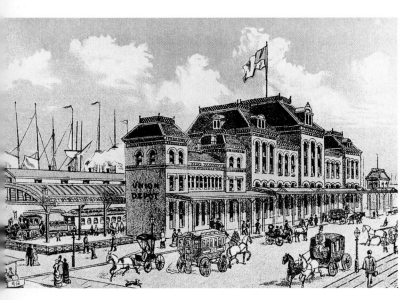

180. Chicago train station (D. Dalziel, *Chicago Town*, n.p., ca. 1884; Newberry Library)

THE ARRIVAL OF THE RAILROAD in Chicago brought great changes to both the city's economy and its physical shape. Canal-era trades such as grain and lumber added rail connections to their networks, while retaining their maritime roots. Other trades such as livestock, package freight, and passenger traffic were wholly reliant on the rails, while the emergent heavy industries became heavy users of both rail and maritime routes. The shape of the city was altered too, as shorelines changed and huge rail yards and industrial districts were created. Once these changes had taken place they could not easily be undone, so that Chicago's plan is still heavily influenced by the nineteenth-century railroad development.

Figure 179 shows some of the early rail lines at the mouth of the Chicago River. On the right-hand side of the mouth is a huge lumberyard and behind that the McCormick reaper works, both of which were served by the Chicago and North Western line on the north bank of the river. On the left-hand side of the mouth were two huge, steam-operated grain elevators (note the chimneys); one of them alone could hold seven hundred thousand bushels of grain.

181. Aerial photograph of the Chicago stockyards in 1938 (University Library, Champaign-Urbana)

182. Map of Chicago showing the railroads about 1866

Tying in to these elevators on the left are the tracks and station of the Illinois Central Railroad, established on this site since the early 1850s. The trains came from the south along a trestle at the water's edge, protected from the lake's waves by a breakwater. This breakwater was the price the city exacted for the Illinois Central's access to the river mouth by way of the lakefront. The tracks divided near the river mouth, with some going into the elevators and others continuing straight into the passenger station, with its long shed and three-storied headhouse.

Headhouses were a characteristic feature of nineteenth-century stations. Figure 180 is a view of the Union Depot, at which a variety of railroads arrived from north and south. The architecture of the headhouse is very characteristic, and so is the train shed seen behind it, with the locomotive at the platform. Behind the shed may be seen the masts of ships in the Chicago River, for the tracks serving the Union Depot ran north-south alongside the river.

Figure 182 shows the railroad system in 1866. Union Station, the Illinois Central station, the Chicago and Northwestern station, and the LaSalle Street passenger station were all clustered in the business heart of the city. The grain elevators, factories, and lumberyards were still tied to the network of water transport, but they were also served by the railroads, which did not destroy the maritime trade but relegated it to the large bulk cargoes. The new element, the stockyards, is conspicuous by its lack of dependence on the water; these yards depended exclusively on rail transport, which could ship their perishable cargoes rapidly and reliably, and were situated alongside the railroads that served them.

In the early 1860s three of the four stockyards were within the city limits. In order to consolidate the operations and remove them from the city proper, in 1865 the Union Stockyard was built on an area that had formerly been a marsh, to the south of the city. From the first, rail transport was a crucial element in this gigantic enterprise, which for the next hundred years would bring Chicago so much wealth and so much notoriety. Figure 181 shows the yards entirely surrounded by railroad lines, which curved in from the west to the great platforms where the beasts were unloaded, and serviced the yards on the other three sides. The South Branch of the Chicago River does indeed stretch to the area of the yards, but it was used more as a sewer than as a means of transport, fitting symbol of the new predominance of the rail system in the city.

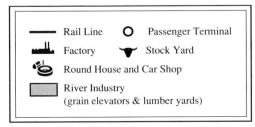

——— Rail Line	○ Passenger Terminal
⚒ Factory	▼ Stock Yard
⊙ Round House and Car Shop	
▨ River Industry (grain elevators & lumber yards)	

183. Photograph of tracks at West Chicago about 1900 (West Chicago Historical Museum)

West Chicago and the Northern Rail Network

As Chicago's economic sphere of influence expanded, its railroads reached far out to the west and northwest, bringing a huge volume of traffic to the city. The area around Chicago came to have an extraordinary density of railroad lines and developed major junctions like Aurora, Joliet, and West Chicago, which were in effect "railroad towns." There were five main groups of lines. The north and west were served by the Chicago and Northwestern and by the Chicago, Burlington, and Quincy. For a time they both funnelled into the Chicago area through West Chicago, using the spectacular three-line junction shown in figure 184. This was the Turner Junction, which later became known as West Chicago; figure 183 is a view of the same junction from the east about 1900.

Up from the southwest came the Chicago, Rock Island, and Pacific and the Chicago and Alton railroads. They both passed through Joliet, where a large roundhouse still survives. Some towns like Joliet harbored the ambition to outflank Chicago on the routes to the West, and so themselves to become great rail centers. In 1871, for instance, Hopkins Rowell published *The Great Resources and Superior Advantages of the City of Joliet,* in which he predicted that "the vast shipments of grain, cattle, etc., from the teeming southwest, over the Chicago, Burlington & Quincy Railroad, instead of paying tribute to the elevators and Stock Yards of Chicago, will be brought here, and shipped east, by the Joliet Cut-

Off." Alas for such hopes: Chicago's advantages in waterborne transport, as well as the indefatigable activity of its politician-boosters, ensured that most lines continued to enter and leave from the city.

From due south came the line of the Illinois Central, passing along the lake into the terminal at the mouth of the river (figure 179). From the southeast came the fourth group of railroads, the Michigan Central and the Indiana Central, while from the East came a variety of lines, some of which were eventually swallowed up by the mighty New York Central. These last lines provided the link to the East.

This immense network in an extraordinarily short period made Chicago the great entrepot for the whole region west to the Rockies. It also had an effect on the immediate countryside, for many of the railroad companies deliberately fostered the growth of suburban communities along their tracks, which then served to carry commuters to their work in Chicago. The Chicago and Northwestern Railroad was particularly successful in establishing such towns; some of them are marked in figure 185. On the line immediately north out of Chicago, a whole string of lakeside commuter communities emerged, and to the northwest towns like Park Ridge, Des Plaines, Mount Prospect, Arlington Heights, and Palatine were called into being largely by the railroad. One could almost say that, as the Illinois Central colonized the middle of the state, so the Chicago and Northwestern colonized northern Cook County.

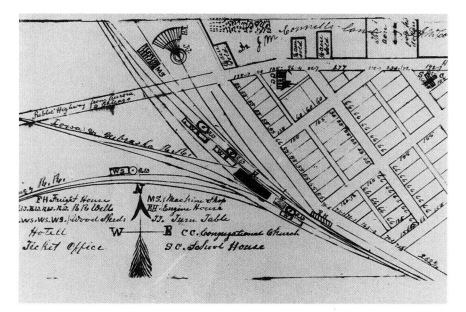

184. Detail from map of West Chicago, about 1860 (DuPage County Atlas manuscript at West Chicago Historical Museum)

185. Aerial perspective of the rail network around Chicago about 1866

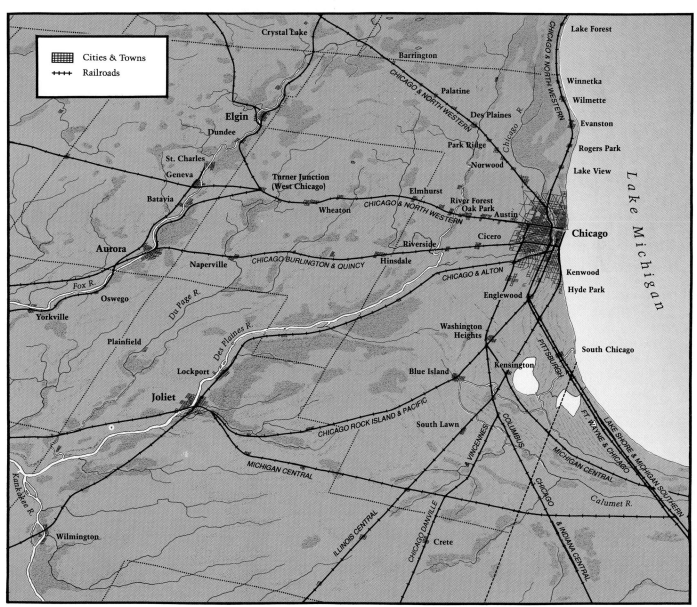

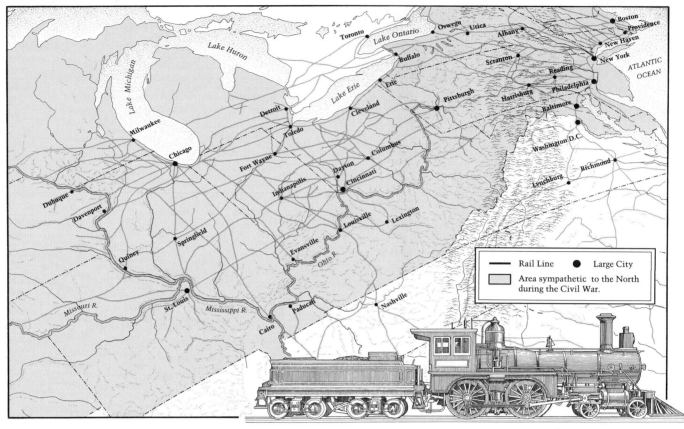

186. Map showing the major east-west rail lines in the 1860s

The East-West Railroads and the Mississippi

IN MANY WAYS, the obvious place for the railroads to cross the Mississippi River on their way west was Saint Louis. It lay at the confluence of the Missouri and Mississippi rivers and by the early nineteenth century was the leading city in the West, boasting a commerce based on the entire Missouri, Mississippi, and Ohio river systems, as well as from New Orleans and the gulf. It was, in short, the "natural" center for north-south trade by water, and as the middle of the nineteenth century approached, Saint Louis's boosters were certain that its dominance would continue and make it the "natural" point for the transcontinental lines to cross the great river—which does appear plausible, when one looks at the map of North America (figure 186).

The confident and well-established merchant patricians of Saint Louis reckoned without the brash

187. (right, above) Eads Bridge under construction (Thomas Cooley, ed., *The American Railway*, Chicago, 1889; Newberry Library)

188. (right) Modern photograph of the Eads Bridge from the riverbank (Illinois Department of Transportation)

energy and inventiveness of the Chicagoans. The opening of the Illinois and Michigan Canal in 1848 and the establishment of telegraphic communication with New York in the same year began to establish a very powerful economic system in which the New York financiers who wanted the western trade came to have a crucial interest in the success of Chicago. When in the same year the Galena and Chicago Union Railroad made its first run, the stage was set for tremendous growth in this east-west connection, whereby Chicago railroads would wrest the midwestern trade away from its old river courses based on Saint Louis.

The leaders in Saint Louis persisted in thinking of the new railroads as mere feeders for the well-established system of riverboats and were taken by surprise when, for instance, the coming of the Illinois Central into southern Illinois diverted the produce of that whole region away, northwards. Saint Louis was unlucky to suffer dislocations during the Civil War, but the city's leaders were also tardy in adopting new technologies. Their railroads were badly built and financed, and the city long opposed the construction of bridges over the Mississippi. It was not until 1865 that the first grain elevator was built in Saint Louis, at a time when Chicago had long possessed at least a dozen large ones.

Meanwhile Chicago's rail network reached out in all directions and by the mid-1860s was ready to cross the Mississippi, to tap the rich Iowa farmland. In 1865 the Chicago and Northwestern Railroad bridged the great river at Clinton, and in 1868 the Chicago, Burlington, and Quincy Railroad crossed it at Burlington; clearly it was time for a bridge at Saint Louis, if that city were not to lose all its trade to the upstart on Lake Michigan. Between 1867 and 1874 Captain Eads was employed in the technically difficult task of building a railroad bridge over the Mississippi at Saint Louis. Figures 187 and 188 show the bridge under construction and completed; note the great construction ties in figure 187, which make the bridge look as if it were to be a cantilever bridge.

Once the bridge was completed, more traffic be-

189. Detail of East Saint Louis (*Atlas of the State of Illinois*, Chicago, 1876; Newberry Library)

gan to flow from the East to Saint Louis. But until then, as figure 189 shows, the railroad lines all had to stop at the edge of the levee (note the single line on the "R.R. Bridge"), and the advantage gained by Chicago would never be recovered. Figure 189 recalls the previous history of the state as well: in the top left is Indian Lake, reminding us that the Cahokia area was a great center of Indian settlement, and the entire street pattern of East Saint Louis shows its origin in the long-lot field pattern of Cahokia.

190. Photograph of the Aurora roundhouse about 1870 (Aurora Historical Museum)

The Roundhouse: Hub of Early Railroad Operations

AURORA WAS THE BIRTHPLACE in 1849 of the Chicago, Burlington, and Quincy Railroad, which at first served Chicago through the West Chicago junction (figure 184). During the 1850s this railroad expanded westwards, with a ferry across the Mississippi River at Burlington; like most of the other Chicago railroads, it prospered enormously during the Civil War and eventually became the basis of the great Burlington Northern line.

The CBQ, as it was familiarly called, maintained a substantial base at Aurora, where its great roundhouse still survives. Our aerial photograph shows it in October 1985, before the restoration that has made part of it into an attractive commuter station and shopping arcade. In 1985 the remains of another roundhouse could be seen to the south, in the outlines of the engine stalls on the cleared ground. Figure 191 shows the existing roundhouse as it would have looked about 1860, when it had about two thousand employees and served about thirteen thousand miles of track.

The heart of the enterprise is the turntable, cranked around by hand, allowing the engines to be backed up into one of the encircling stalls; figure 193 shows the interior of the Aurora roundhouse.

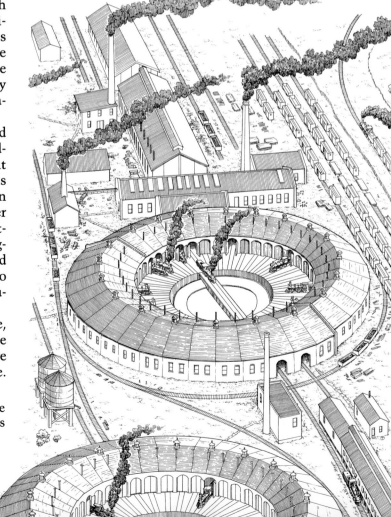

191. Aerial perspective of the main roundhouse in operation during the 1870s

192. Aerial photograph of the roundhouse at Aurora, October 1985

Steam engines had to be cleaned out at the end of each day's running, when the ashpans were emptied into a pit beneath the tracks and a thorough inspection was carried out. At the same time the exterior of the engine was thoroughly cleaned, since nineteenth-century engineers took pride in the spotless appearance of their machines. If there was a major problem, the engine could be taken from its stall into the locomotive erecting house, usually known as the "back shop," where it could if necessary be entirely rebuilt. There huge travelling cranes were mounted in the roof, and furnaces and lathes allowed metalworking of any part. When the engine was ready, it left the roundhouse by the south side, where there were coaling towers and watertanks. Of course, the problems of ash clearance, watering, and coaling have disappeared since the advent of the diesel locomotive, and roundhouses have been replaced by much less distinctive long sheds.

193. Interior of the Aurora roundhouse, about 1870 (Aurora Historical Museum)

Stickney: Classification Yards Ancient and Modern

THE GREAT INCREASE IN RAIL TRAFFIC during the nineteenth century led to problems in sorting out the freight cars. Nowhere was this problem more acute than in Chicago, where in the winter of 1909–10 congestion was such that the entire system came to a standstill. By then, various methods for speedy classification were being tried. Part of the solution lay in removing as many yards as possible from the congested area of downtown Chicago, and part in redesigning the yards themselves.

One of the most ingenious ideas was proposed in 1889 by A. B. Stickney, of the Chicago Union Transfer. He envisaged a huge circular track on which trains would run counterclockwise, shunting off their cars at a series of points (whether inside or outside the circle is not quite clear). Figure 194 shows what this circle looked like, built out on the edge of

the metropolitan area and well away from the congested yards just south of the Loop. Alas, Stickney's project never worked well, partly for technical reasons and partly because it was very hard to persuade the competing railroads—led by strong personalities like Mr. Vanderbilt—to cooperate in a collaborative operation. Until recently it was possible to see from the air the remains of Stickney's circle, to the south of the new yards (figure 196); these remains seem to have been obliterated by a new classification yard for containerized goods.

By 1900 it was clear that some other form of yard was needed. In the one-way hump yard the cars were pushed to the top of an incline and then rolled down onto an appropriate track; in 1902 this was changed to a two-way hump yard, unique for its time and surviving in its essentials to the present day. This great

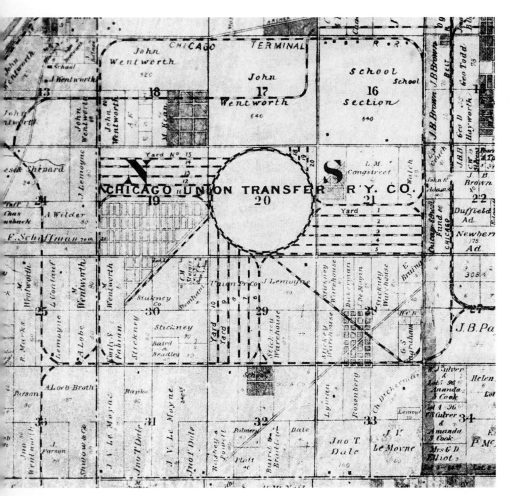

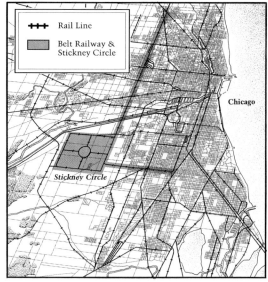

194. Stickney Circle (*Snyder's Real Estate Map of Cook, DuPage, and Part of Will Counties*, Chicago, 1898; Newberry Library)

195. Map showing the Stickney Circle and the Chicago railroads in the 1890s

196. Aerial photograph of the Belt Railway of Chicago classification yard, about 1970 (University Library, Champaign-Urbana)

classification yard came to be operated by the Belt Railway of Chicago (BRC) and in its early days employed many casual laborers. Their task, manually to control the cars as they rolled down the slope and onto the back of the next car, was difficult and dangerous, for brakes could be uncertain. Sometimes the workers would deliberately engineer "blow-ups" (hard collisions), however, and would pilfer from any cars that might be broken open. These days the whole operation is automated, and up to six cars a minute can go over the hump in each direction.

Twelve railroads jointly operate the BRC, thus making it a "neutral" area for customers. This feature of the operation led in the 1920s to the development of the Clearing Industrial District, an area located on the boundary of the city, just north of the yard, where industrialists were encouraged to establish factories, secure in the knowledge that they could get highly competitive freight rates for their supplies and products. This "city of factories" was an early example of large-scale collaborative planning and became an influential model for other industrial communities.

197. Photograph of the Clearing Industrial District logo

198. Aerial photograph of winter scene with abandoned track in north central Illinois, April 1988

Vanished Railroads in the Illinois Landscape

URING THE TWENTIETH CENTURY the role of the railroads was substantially diminished by road haulers, who ate into the freight market, and by airlines and private automobiles, which combined to greatly reduce the passenger market. The result was the abandonment of many railroad lines, but this has not meant their effacement from the landscape. In many cases towns were aligned with the railroad, as were some field boundaries out on the former prairie.

Flying over Illinois we constantly notice patterns imposed by vanished railroad lines. Figure 198 is a photograph taken in April 1988, when a light snow was covering abandoned tracks near Byron, in northern Illinois. The track comes down from the middle top of the photograph, where it is just possible to see at points the outline left by the railroad ties. At the creek the bridge has totally disappeared; it must have been made of wood, as stone abutments gener-

ally survive and often give useful confirmation of a railroad's former existence. The track on the near side of the river still seems to be in some kind of use, for the bed has been deepened and the tie marks totally effaced; perhaps people drive down there for the fishing.

The other photograph was taken near Bishop Hill in October 1988. It too shows part of an abandoned line, coming in diagonally from the top left. The trees mark the line's former alignment, even though the roadbed is no more than a rutted farm track. Note the grain elevator, which seems largely abandoned, and the buildings alongside the track, one of which may have been the depot. This little cluster may once have looked rather like the scene in Figure 177. A worthy project for a local university or historical society would be to identify an abandoned track of this kind and then to study the former railroad structures that could be found along its path.

199. Aerial photograph of abandoned railroad with grain elevator near Bishop Hill, October 1988

9

CHICAGO BECOMES THE SECOND CITY

Introduction: Chicago from the 1850s to the 1880s

FIGURE 200 SHOWS CHICAGO towards the end of the 1860s, after a decade of growth which had roughly tripled its population, to about three hundred thousand people. The Civil War had been enormously profitable to Chicago's merchants and manufacturers, but even without that stimulus there would have been astonishing growth. We can see in the figure the original sources of the city's prosperity: the Illinois and Michigan Canal stretch-

200. View of Chicago (*Harper's Weekly*, 1871; Chicago Historical Society)

note for instance the roundhouse just inside the large bend on the South Branch of the river. Note, too, away to the south, the Union Stockyards, whose very existence at that site and on that scale was due to the railroads.

The other, newer elements in Chicago's economic system do not emerge so easily in this view. West of the South Branch are many iron and brass foundries, laying the foundation of Chicago's preeminence as a manufacturing center. In the business district are merchandizing enterprises, from which would emerge the huge mail-order business. Finally, there was the whole system of banking and commodities exchange, dependent since the late 1840s on swift telegraphic communication with New York and so with the whole network of international capitalism.

Looking at the figure, we have the image of a well-defined central area, with substantial buildings. Beyond that the artist gave a mere impression of structures, which corresponded with reality, for the city was growing so fast that always at its fringes was a layer of hastily constructed wooden shanties, in which the most recent immigrants usually lived. This layer was rightly regarded as a hazard from the point of view of fire, disease, and social disorder.

Fires were frequent and culminated in the disastrous conflagration of 1871. Disease was also endemic; the houses often lay low in the muddy prairie, with little possibility of adequate drainage, and the water sources were often polluted with a variety of microorganisms that spared no one, not even the rich. It would have been surprising, too, if social discontent had not been rife in so large a laboring population, particularly since it contained many politically conscious recent immigrants from countries like Germany, where ideas of social justice and socialism were widely held.

The response to this discontent took many forms. Some employers adopted what they regarded as a benevolent paternalism; of these, George Pullman is a prime example. Others in the society looked to direct social action. The churches were constantly prominent in this effort and were joined by wellborn women like Jane Addams, whose experiment at Hull-House became famous throughout the world. Others thought more in terms of direct repression, using the federal force that was available after the establishment of Fort Sheridan. One way and another, Chicago avoided the social explosion that its social and economic conditions seemed sure to bring about.

ing away to the southwest, the immense lumber-yards (some of them in the canal's terminus slips), the grain elevators, of which the most spectacular are at the river mouth, and the ships on the lake.

We can identify the railroad system, which brought most of the Midwest into Chicago's economic orbit. The Illinois Central tracks and station at the river mouth are the most obvious parts of this system, but others can be picked out in the city;

201. *Chicago in Flammen* (Eugen Seeger and Eduard Schlaeger, *Chicago: Entwicklung, Zerstörung, und Wiederaufbau der Wunderstadt*, Chicago, 1872; Newberry Library)

The Great Fire of 1871

OST CITIES HAVE HAD THEIR "GREAT FIRES," but the one that broke out in Chicago in October 1871 was more destructive than most. The summer had been very dry, so that both the city and the surrounding countryside were parched. During the first week of October several fires started but were contained, and the apprehension of fires was such that businessmen deliberately kept their goods out of areas of the city regarded as particularly vulnerable.

One of these areas was the stretch of wooden shanties, homes, and businesses that lay southwest of the central area, down towards the stockyards. There, on the night of October 8, began a fire that, fanned by a southwest breeze and at first neglected by the exhausted firemen, soon became uncontrollable. Figure 203 gives us some idea of the immense area that eventually burned; the fire jumped twice over the river in the process. Chicago was stuffed at that time with combustible materials, often housed in equally combustible wooden houses and sheds. Once the fire was well under way, it seems to have induced the type of fire storm to which some German cities were subjected through aerial bombing during the Second World War; the rising heat was of such a degree that it created a column of ascending air, which had to be replenished at its base by drawing in fresh supplies of air.

Many of the eyewitnesses testify to this great wind, stronger by far than the unaided southwest breeze. It would lift up whole sections of burning roof—often made of tarred felt—and whirl them onto the roofs of other houses hundreds of yards away. The people of Chicago and the firemen could not fight the fire on any single front, for it constantly broke out where it was least expected, and often the flames had, as witnesses put it, the force of a blowtorch. In the end it was the rain that extinguished the flames, twenty-seven hours after they had broken out.

During the fire there were examples of great heroism, and of violent thievery; as one contemporary

202. "Scene at the Junction of the Chicago River" (E. J. Goodspeed, *History of the Great Fires in Chicago and the West*, New York, 1871; Newberry Library)

203. Aerial photograph of the area of the fire, about 1975 (Seinwill Collection, Newberry Library)

204. View of the devastated area (Eugen Seeger and Eduard Schlaeger, *Chicago: Entwicklung, Zerstörung, und Wiederaufbau der Wunderstadt*, Chicago, 1872; Newberry Library)

put it, "It takes all sorts of people to make a great fire." Chicago was the headquarters of Civil War hero General Philip Sheridan, and he soon had about seven hundred troops on the streets, helping to keep order. Although the destroyed area was extensive, many important installations were not harmed; the Union Stockyards were intact, as were most of the grain elevators, lumberyards, and factories, as well as at least one railroad depot. Moreover, eastern financiers had a vital stake in the continuing success of the city, and they willingly put up money for its reconstruction. So Chicago, helped by contributions from all over the world, was rapidly able to reestablish itself.

The Chicago Fire was part of a larger catastrophe that took place that summer. The woods of Wisconsin and of Michigan were dry and in many places full of debris left by timber cutters. At about the same

time as the Chicago Fire, a conflagration developed in the central Wisconsin woods and raged under the southwesterly wind until virtually all the eastern part of the state (a 150-mile-wide swath) had burnt. The fire was so fierce that trees on islands half a mile out in Little Sturgeon Bay were ignited, and the loss of life was appalling. Peshtigo, a lumber town of about two thousand people, was the worst hit, losing about six hundred of its inhabitants, roughly twice as many as perished at Chicago. The whole catastrophic chain of events was connected, for the northern forests were heedlessly cut over, leaving vulnerable trimmings, in order to construct the wooden houses that burned so well in Chicago. Still, fire services the world over learned from the Chicago experience, and precautions were taken in planning and building to make such a disaster less likely in the future.

CHICAGO BECOMES THE SECOND CITY · 149

205. Aerial photograph of Stearns Quarry, April 1986

Rebuilding after the Fire

ONE OF THE INCONSPICUOUS ADVANTAGES of Chicago's site was that it lay adjacent to enormous beds of limestone and so became surrounded by quarries. Figure 208 shows the sites of these and of the clay pits suitable for brickmaking. Figure 205 is an aerial view of one of the largest of the limestone quarries, just south of the I-55 highway and a little farther south of the eastern terminus of the Illinois and Michigan Canal. Indeed, one of the canal slips may just be seen in the top right-hand corner of the photograph. The disused quarry is very deep, as may be judged from the size of some of the large trucks at the top of it, where they seem to be dumping trash down its sides.

The limestone that came out of this quarry was not suitable for facing-stones or even for sidewalks, but it could provide satisfactory footings for buildings and roads; no doubt the quarry was in heavy use after the fire. For superior-quality limestone, build-

ers looked to the Joliet area, where the so-called Athenian marble was quarried, as shown in figure 206. Here the blocks of stone are being carefully cut from the face and lifted with a sheerlegs onto a special low-loader cart, which is taking them to be eased onto the railroad cars at the bottom of the engraving. This Lemont stone was a creamy-white color and, when used as facing-stone, made many of Chicago's buildings of the 1850s very attractive. It was also used as flagstones for sidewalks, which during the fire are said to have "sizzled and burned."

The rebuilding of Chicago went on at a tremendous pace after the fire, mostly using the styles that were already familiar, and indeed often flouting the new regulations concerning fireproofing. One of the most characteristically handsome of the new buildings was the Chamber of Commerce (figure 207); this huge edifice opened for occupancy on the first anniversary of the fire.

206. Detail from the
*Combination Atlas Map
of Will County* (Elgin, 1873;
Newberry Library)

■	Area of the Great Fire
■	Clay (Brick-making)
✗ Chicago Limestone	✗ Clay Pit
✿ Glacial Boulders (Ornamental Stone)	
→ Indiana & Joliet Limestone	

208. Map showing building
resources around Chicago

207. Contemporary photograph of the rebuilt Chamber
of Commerce (Chicago Historical Society)

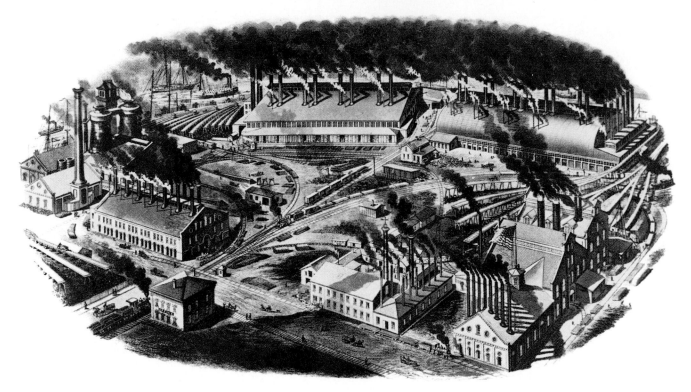

209. Bird's-eye view of the North Chicago Rolling Mill Company factory, nineteenth century (Newberry Library)

The Dark Satanic Mills of Chicago

210. Advertisement for the Franklin Iron Works, 1867 (Newberry Library)

O F THE MANY FACTORIES that emerged and thrived in Chicago during the Civil War, few were more spectacular in their appearance than the iron foundries, with their stupendously belching smokestacks. The oldest of these enterprises was the rolling mill, founded in 1857, which became the North Chicago Rolling Mill Company (NCRMC). In 1865 it produced the first steel rails made in the United States and in 1869 built two blast furnaces, just a year after Chicago's first furnaces, constructed by the rival Chicago Iron Company. In 1872 the NCRMC adopted the Bessemer process, and by 1876 Chicago accounted for one-third of all rail production in the country.

Figure 209 shows the NCRMC factory towards the end of the century. In the background are the ships that brought the ore from the upper Great Lakes, and in the foreground is one of the trains that contributed the coal from Indiana and Illinois. Limestone was also available locally, so that all the elements were at hand to shape a product to serve the Midwest market. The many small chimneys service individual foundries in the sheds below them; two huge furnace chimneys may be seen on the extreme left, just to the right of the very tall chimney. This works probably lay a little way up the North

Branch of the Chicago River. Many other metal-working factories soon developed, like the Franklin Iron Works (figure 210).

The large-scale wholesale trade was concentrated in the downtown area, on South Water Street (Figure 212); note the ships in the river, ready to unload their cargoes into the warehouses. As the railroads largely replaced the river traffic, merchandizing migrated to State Street, better sited for the new mode of transport. The central area soon became too congested for heavy manufacturing, and from about 1880 onwards many foundries and mills moved to a new industrial area by Lake Calumet.

At Lake Calumet a barren marshy area, with virtually no settlement, held the possibility of easy land reclamation in the sandy soil and ready access to the pure water of the lake. More and more steel mills were established in the Calumet area, and the Chicago industry declined. Between 1872 and 1892 ore imports increased fifteenfold, as the Calumet area became one of the nation's great steel-producing areas, eventually rivalling Pittsburgh itself.

211. Detail from a bird's-eye view of Chicago, about 1892 (Currier and Ives, New York; Newberry Library)

212. View of South Water Street (Otto Jevne and Peter Almini, *Chicago Illustrated*, Chicago, 1866–67; Newberry Library)

213. View of the Chicago River (Otto Jevne and Peter Almini, *Chicago Illustrated*, Chicago, 1866–67; Newberry Library)

River Pollution and the Lake Cribs

THE CHICAGO RIVER WAS SMALL and sluggish. As the city used it for progressively greater quantities of effluent from tanneries, distilleries, and slaughterhouses, it became very heavily polluted. Operators of tugboats like the ones shown in figure 213 had great difficulty in using the river water for their engines, so full was it of foreign bodies. In times of heavy rainfall, this pollution could drift quite far out into the lake, and at the best of times it was enough to contaminate the city's water supply, which was drawn from the lake just north of the river mouth. This pollution surely contributed

214. Diagram of the operation of the new water system (*The Tunnels and Water System of Chicago*, Chicago, 1874; Newberry Library)

to the epidemics of cholera and typhoid that ravaged the city during the 1850s and 1860s.

Deciding that a radical solution was needed, the city council in 1863 empowered the city engineer to study means of drawing water from uncontaminated parts of the lake, and the idea of a deep tunnel two miles long was born. It was an audacious concept; as a pamphlet of the time observed, "New York declared it could not be done; New England affirmed, in Puritanic style, that it should not be." Nevertheless, between March 1864 and November 1866 it *was* done. The "crib," which was "almost as large as the Palmer House," was floated out to the chosen spot two miles off shore and there sunk. Meanwhile a tunnel was dug from the shore and another from

the crib; they met with only an inch of difference, an astonishing achievement in the days before laser technology. The crib and tunnel were always shown to important visitors and were indeed remarkable engineering feats for their time.

Figure 214 shows how the system operated. Water was drawn from the bottom of the lake, two miles out, and flowed down the gently (four feet in two miles) sloping tunnels (a second tunnel was added in 1872) to the new pumping station. Thence it was pumped to the top of the new watertower—one of the few buildings to survive the fire of 1871—and so distributed smoothly into the city's water mains. This system was a great improvement, and other tunnels and cribs followed over the years. Figure 216 shows a crib as it appeared circa 1866, and figure 215 not only shows a crib as it now is, but also gives a good impression of Chicago as a waterfront city, an impression that is hard to get without going out some distance on the lake.

215. Aerial photograph of one of the cribs, April 1986

216. View of one of the cribs (*Chicago Illustrated*, about 1866; Newberry Library)

217. Aerial photograph of the Glessner House and part of Prairie Avenue from the northeast, April 1986

Residences of the Rich

218. Photograph of Prairie Avenue, nineteenth century (Chicago Architecture Foundation)

In Chicago's early days the rich usually lived in the city, whose air was not yet so polluted that suburban flight seemed desirable. One very fashionable district that developed after the fire of 1871 was on Prairie Avenue, particularly between Sixteenth and Twenty-second streets (figure 218). Here the great industrial magnates like Philip Armour and George Pullman built their townhouses, among the factories and railroad lines.

Eventually, as figure 217 shows, the area was overwhelmed by industry, and by the 1930s almost all the Prairie Avenue houses had been abandoned by the original families. A small section of the avenue has been preserved, however, with a few of the original houses. Undoubtedly the most unusual of these buildings is the Glessner House, built in 1886 for J. J. Glessner, a founder of the International Harvester Company. It is now the home of the Chicago Architecture Foundation, which runs tours and programs there.

The house may be seen occupying the lower right corner of the grassy block in our figure; its high walls enclose a courtyard. Most of the houses on the avenue were like the two to the east, facing Glessner House: freestanding buildings in light, airy styles like the so-called French chateau. Glessner House, on the other hand, aroused much contemporary criticism because of its fortresslike Romanesque appearance from the street. All the main rooms faced

219. Aerial photograph of the McCormick Mansion at Cantigny, April 1986

onto the inner courtyard, rather like the old town-houses of some cities in Spain and Italy. Glessner and his architect, Henry Richardson, knew exactly how different their house was and may have been influenced by fear of urban tumult, Glessner and his wife having experienced at first hand the Haymarket Riot of 1886.

Like many other rich Chicagoans, the Glessners also had a home in another part of the country, in their case in New Hampshire. Chicago differs in this respect from the great cities of the Old World, surrounded as they often are by the country mansions of the rich. When those mansions were built, it was often quite difficult to have a secondary residence far from the city. With the coming of the train, all that changed, so that the Chicago millionaires could have a second residence almost anywhere in the United States, from Florida to the Canadian border.

The countryside around Chicago thus lacks substantial country houses, with one or two exceptions of which Cantigny (figure 219) is the most notable. This mansion, set in superb grounds, was the country home of Colonel Robert McCormick, who named it after a village in France where he had served in 1918 with the U.S. First Division. During the house's most recent renovation, in the 1930s, McCormick specifically charged the architect to create a building that might "become one of the show places of Chicagoland because of its sheer beauty." Today, this wish has largely come true; the magnificent house and gardens are open to visitors, as is the adjacent First Division Museum.

220. Aerial photograph of the neighborhood north of Saint Paul's Church, about 1930 (Chicago Historical Society)

Dwellings of the Poor

BY THE 1880S THOUSANDS OF EUROPEANS were arriving in Chicago to seek employment in the many factories that had developed along the North and South branches of the Chicago River and that were spreading constantly westwards. These workmen usually found housing as cheap and as close to their factories as possible. Consequently the industrial area was interspersed with tracts of substandard housing, like the tenements above.

This aerial view of about 1930 shows the housing about two miles north of Bridgeport, the terminus of the Illinois and Michigan Canal. We are looking to the south, towards the International Harvester plant. The houses are very close-packed—there were often as many as one hundred people per acre in these areas—and are mostly of balloon-frame construction. Many of them probably date from before the fire of 1871, for this area lay safely to the west of those flames.

Figures 221 and 222 offer a closer look at such an area, a little way to the north of this one. We are able to reconstruct the block quite accurately, because it was delineated not only on the Hull-House nationalities map but also in Robert Hunter's work, *Tenement Conditions in Chicago* (Chicago, 1901). On the map it is the fourth block down from the top, on the left-hand side, bordered on the north by Taylor Street and on the south by De Koven Street. There is a considerable mixture of nationalities, with Italians and Bohemians predominating. On this block,

as on several of the others, a middle alley has been "opened by ordinance."

The perspective view shows the same block from the Taylor Street side. At street level there are shops, as well as a stable and a church; above the shops and saloon is the housing. The buildings were densely packed together, and the rooms inside very crowded. Large immigrant families lived in three or four rooms, which were ill-lit and badly ventilated. The alley did let a little more air into the block, but it had no sewage system, relying on outhouses bordering the alley. To this insanitary arrangement was added the refuse generated by the many stables, one of which is seen here. Small wonder that diseases of all kinds ran rife among these communities.

City hall made sporadic attempts at reform after the worst epidemics, and eventually there were social centers like Hull-House. From the start the churches towered impressively above the people's houses, as they did in the European cities from which the immigrants came (figure 220). The churches were a great source of support, both material and moral, and formed the focus of most immigrant communities. But the church leaders, no doubt correctly, did not make it their task to undertake those political reforms that might have changed the whole system, and reform efforts from other groups were unavailing. Chicago continued to have large areas of substandard housing, a problem that remains unresolved to this day.

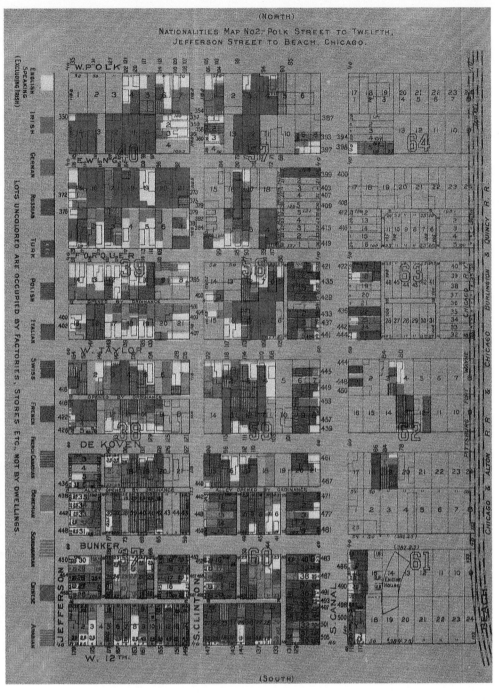

NATIONALITIES MAP No2-POLK STREET TO TWELFTH,
JEFFERSON STREET TO BEACH, CHICAGO.

(NORTH)

ENGLISH SPEAKING (EXCLUDING IRISH)
IRISH
GERMAN
RUSSIAN
TURK
POLISH
ITALIAN
SWISS
FRENCH
FRENCH-CANADIAN
BOHEMIAN
SCANDINAVIAN
CHINESE
ARABIAN

LOTS UNCOLORED ARE OCCUPIED BY FACTORIES, STORES ETC. NOT BY DWELLINGS.

W. POLK
E. W. LING
FORQUER
W. TAYLOR
DE KOVEN
BUNKER
JEFFERSON
S. CLINTON
S. CANAL
BEACH
CHICAGO & ALTON R.R.
PITTSBURG, FORT WAYNE & CHICAGO R.R.
BURLINGTON
CHICAGO, BURLINGTON & QUINCY R.R.

W. 12TH.

(SOUTH)

221. Nationalities map
(*Hull-House Maps and Papers*,
New York, 1895; Newberry
Library)

223. Map showing Chicago's
neighborhoods in the 1890s

Industrial Area ▢ Area of Figure 221
Densely Populated Working Class Wards

122. Aerial perspective of
tenement block

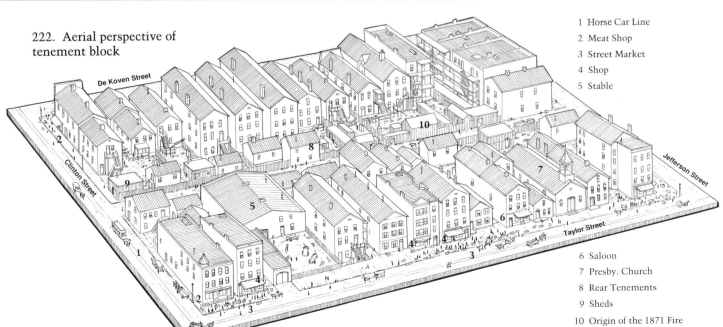

De Koven Street
Clinton Street
Jefferson Street
Taylor Street

1 Horse Car Line
2 Meat Shop
3 Street Market
4 Shop
5 Stable
6 Saloon
7 Presby. Church
8 Rear Tenements
9 Sheds
10 Origin of the 1871 Fire

224. Photograph of Hull-House at its greatest extent (Chicago Historical Society)

Jane Addams and Hull-House

JANE ADDAMS WAS BORN near Rockford in 1860, the daughter of a prominent banker and miller who in his district had been one of the principal sponsors of the Galena and Chicago Union Railroad. She was educated at Rockford Seminary, a liberal arts school for young women, and at home enjoyed a lively social and cultural life with her ambitious and cultivated stepmother. However, like so many of her generation, the first in which women often received higher education, she felt somewhat at a loss after graduation. She knew that she could and certainly should undertake some worthwhile work, but what?

Visits to Europe in 1883–85 and 1887–88 began to clarify Addams's ideas, for while in London she went to the university settlement at Toynbee Hall, where graduates of the English universities ran a sort of social center for the slum dwellers of East Lon-

225. Photograph of Jane Addams (Chicago Historical Society)

226. Aerial photograph of Hull-House, April 1986

don. Returning to Chicago, Addams and her companion, Ellen Starr, began looking among the immigrant tenements for a "big house" that might be the Toynbee Hall of Chicago. They found what they were looking for at 335 South Halsted, where the country house built in 1856 by Charles Hull now lay in the middle of the tenement houses of an immigrant community. Addams and Starr moved in and developed an astonishing range of activities for their neighbors: kindergartens, daycare centers, gymnasiums, reading centers, classes in painting and pottery, a choir, a band, ethnic theater, cooking classes, and so forth. To house all this activity, Hull-House was constantly expanding, until it eventually comprised thirteen substantial buildings (figure 224).

Addams had gone into this activity more or less unguided, simply offering what seemed to her to be needed "to provide a center for a higher civic and social life," but as time went by she found that she had remarkable talents for fund-raising and for self-promotion. Some of her friends and helpers, particularly Florence Kelley, encouraged the development of social studies at Hull-House, where the famous "nationalities maps" (figure 221) were drawn up as an aid to helping the local people resolve their housing problems. All these friends and helpers lived at Hull-House whenever possible, thus getting an intimate acquaintance with the neighborhood's problems.

Addams was inevitably drawn into political activity, becoming an early supporter of unions, of women's suffrage, and of international pacifism. These were widely regarded as dangerous causes, and some of her well-heeled patrons began to back off from the Hull-House experiment. For some time during and after the First World War, her reputation underwent a decline, but towards 1930 it began to revive, and when she died in 1935, many people considered her a sort of lay saint, a Mother Cabrini without the habit. With the passage of time, she seems both more prescient and less selfless than her earlier image. The causes that she espoused have largely been accepted, but her way of running Hull-House has come to seem rather self-centered and maternalistic. Still, the institution survived and flourished after her death, decentralizing its operations in the city when Hull-House itself became part of the area appropriated by the University of Illinois in 1963. Figure 226 is a recent aerial view of Hull-House, showing the two buildings restored (and one resited) by the university, but now dwarfed by their proximity to the brutal concrete blocks behind them.

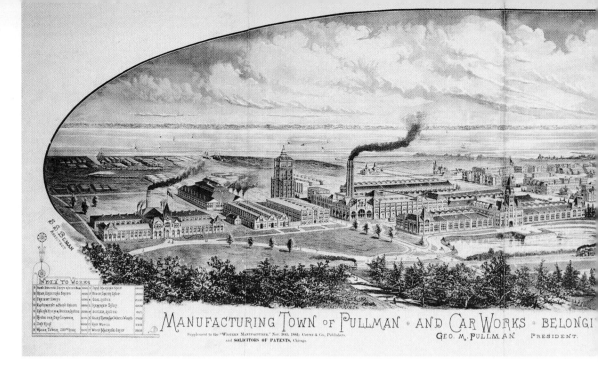

227. Bird's-eye view of Pullman, nineteenth century (Chicago Historical Society)

An Experiment in Paternalism: Pullman

THE EARLY 1880S SAW A HUGE NEW DEVELOPMENT of industry in the area south of Chicago; the Pullman venture was part of this. George Pullman came to Chicago from the East in 1855 and was at first known chiefly for his skill in moving houses (both vertically and horizontally), his most celebrated feat being the six-foot elevation of the huge Tremont Hotel, without any interruption in its services. From the late 1850s he turned his attention to the problem of train travel, for journeys could take several days, and the cars of the time were so rough that people arrived exhausted. In 1864 Pullman built a new type of sleeping carriage, longer, wider, higher, and heavier than any hitherto constructed. It was also five times as expensive and could not be used without alterations to most existing rights-of-way. Still, the need for greater comfort was such that these new cars caught on, and after the establishment of Pullman's Palace Car Company in 1867, they began to be used nationwide.

To sleeping cars Pullman soon added restaurant cars, built to the same high standards, and these too were a huge success. He retained control of their operation by agreement with the railroads and was soon renowned for his service overseas as well as in the United States. Needing more space to build his cars than was available in Chicago, and wishing to avoid the labor unrest that marked the 1870s there, in 1880 he bought four thousand acres of marshland by Lake Calumet (figure 230) and built not only a

huge factory but also a model town (figure 227, looking eastwards with Lake Calumet in the background). This had housing appropriate for each level of company employees and a church, hotel, library, market, school, and theater. Pullman did not allow employees to buy their houses, for as he explained, "If I had sold my sites to my workmen at the beginning of the experiment, I should have run the risk of seeing families settle who are not sufficiently accustomed to the habits which I wish to develop in the

229. Aerial photograph of Pullman, April 1986

228. Interior of the car works at Pullman (*Chicago: The Great Central Market*, Chicago, 1921; Newberry Library)

inhabitants of Pullman city." At first the experiment seemed to work well, though from the start Pullman's paternalism jarred many, and as nobody could buy a house, "nobody regards Pullman as a real home."

In 1894 the Pullman workers went on strike, and the whole system unravelled. Soldiers of the Illinois National Guard were quartered on the grounds of the Florence Hotel, and the strike was crushed. Pullman was not at all vindictive, and the factory soon thrived again, but the town was doomed as a corporate entity and was gradually sold off. Pullman died in 1897, distressed by the failure of his experiment. Perhaps the best commentary on it came from Jane Addams, who described Pullman as a "modern king Lear," whose overbearing certainty that he knew what was right eventually turned his "children" against him.

Much of the residential area of Pullman survives

230. Pullman area from the USGS 7.5-minute series, Calumet quadrangle, surveyed 1896–97

today (figure 229). This view, looking westwards, shows the central square where all market activity was concentrated, and the common church beyond it. It also shows the rectangular street pattern, which was rather dull, and the generally uniform and harmonious style of the houses. The majority are solidly built of brick, at a time when most workers' houses were hastily run up with wood. Pullman is a fascinating monument to a remarkable experiment, from which different people draw very different lessons.

CHICAGO BECOMES THE SECOND CITY · 163

Another Means of Social Control: Fort Sheridan

DURING THE CHICAGO FIRE federal troops helped to keep order; they were also called in during the railroad strikes of 1877. The Haymarket Riot of 1886 suggested to some of Chicago's leaders that it would be comforting to have a large force of soldiers permanently at hand, and so in 1887 the Commercial Club of Chicago donated to the federal government a large tract of land for a military post by the lake about twenty-five miles north of the city. In 1888 the post was named after General Philip Sheridan, who had recently died, and construction of buildings went forward under the direction of the leading Chicago architectural firm of Holabird and Roche, for there was at the time no uniform rule about how such military bases should be planned and built.

By the end of the century Fort Sheridan had reached the stage shown in figure 232, a contemporary photograph probably taken from a balloon. The tower on the central range is the most conspicuous feature, with a full range of ancillary buildings clustered round it: stables, barracks, hospitals for men and animals, a fire station, a magazine, and so forth. Roads enclose a large central area used for cavalry maneuvers, and outside the roads are the officers' quarters, built of a distinctive yellow brick. At the time, Fort Sheridan stood by itself; there seems to be no trace of buildings to the south of it, where Highland Park now lies.

The troops at Fort Sheridan were used to quell labor unrest in 1893–94 (figure 233). After that, however, the base became chiefly an administrative center. Great numbers of recruits were trained there during the First and the Second World War, and many officers who later became famous served there, including General Patton, who after graduating from West Point in 1909 lived at Fort Sheridan in one of the family quarters.

Over the years additions were made to the original structures, and some parts of the base have assumed new uses: the cavalry ground, for instance, now makes a fine golf course. But the old buildings are almost all intact and, as figure 231 shows, are in some cases very impressive. The central tower is of superb brick workmanship, though it is a pity that for structural reasons it had to be truncated in 1949 (compare its appearance in figures 231 and 232). As of 1989 it seemed probable that Fort Sheridan would be abandoned as a base. But we have to hope that at least parts of it will continue to be preserved as a historic site, redolent of a certain period of the history of the U.S. Army.

231. (opposite, top)
Aerial photograph of the central buildings at Fort Sheridan, October 1985

232. (opposite, bottom) Aerial photograph of Fort Sheridan about 1900 (Fort Sheridan Museum)

233. Contemporary drawing of federal soldiers confronting striking workers (*Harper's Weekly*, 1894; Newberry Library)

10

THE MATURE CITY
AND ITS
INSTITUTIONS

Introduction: The Loop in 1898

THIS EXTRAORDINARY IMAGE of the business district of Chicago contains fascinating elements of survival and of innovation. To take the survivals first, the Illinois Central tracks still cut the center of the city off from the lakefront, which had virtually no docking facilities for pleasure craft. The tracks for two other large stations entered the city from the south, while on the north bank of the river stood the huge station of the Chicago and Northwestern Railroad, roughly where the Merchandise Mart now is.

Michigan Avenue still ended on the south bank of the river, the nearest access northwards being over the Rush Street swing bridge. The swing bridges themselves were notable survivors from the 1860s, with none of their cantilevered successors in sight; it is easy to see how the swing bridges, which needed a substantial "island" on which to stand, impeded navigation on the river. Finally, numbers of elevators were still in the scene, including the (barely visible) elevator at the harbor mouth, which was in some sense Chicago's symbol for many years.

Many changes are also visible; perhaps the most notable was the arrival of the elevated railroad system, the famous "El," which makes the circuit defining the Loop. The El-stations are particularly prominent, and survive virtually unchanged today. Looking carefully, we can also see a dense pattern of streetcars, which made travel to the closer suburbs quick and easy; there were as yet no automobiles.

Some familiar buildings like the Chicago Public Library and the Art Institute had appeared, though the Art Institute was housed in a relatively small building, hemmed in by the tracks. In the city itself some buildings were even taller than the elevators, for this was the first age of Chicago's skyscrapers. The view caught Chicago at a distinctive moment, a sort of breathing space when the center was no longer industrial and was not yet overrun by the automobile. This "streetcar city," as some historians like to call it, may be the most livable urban form so far invented.

234. *Bird's-Eye View of the Business District of Chicago* (Chicago, 1898; Chicago Historical Society)

235. Plan of Riverside,
nineteenth century
(Riverside Historical
Commission)

Emerging Suburbs: The Example of Riverside

BECAUSE THE WHOLE CHICAGO AREA had been surveyed on a quadrilateral system and roads and property lines usually followed these survey lines, the tendency was for Chicago to grow simply by pushing out square block after square block, with an often monotonous effect. From the middle of the nineteenth century onwards some architects and planners protested against this type of expansion, but in general the convenience of the system was such that it everywhere prevailed.

Nine miles west of Chicago, however, was a sixteen-hundred-acre site of unusual interest, gently rolling country by the side of the Des Plaines River. In 1868 the owners of this tract, the Riverside Improvement Company, called upon the landscape architect Frederick Law Olmsted to design a park on this site. Olmsted was then forty-six years old, in partnership with the Englishman Calvert Vaux, and at the height of his powers, having designed and cre-

ated Central Park in New York. His early life was rather purposeless, including stints as a sailor and as a farmer, but by the 1860s he had travelled widely, in Europe and in the United States, and had fully developed his ideas about how small towns should be laid out.

For the Riverside site, Olmsted thought that a town would be more appropriate than a park and that it ought to conform to three main principles: the roads should be elegantly curved, following the terrain as far as possible; the houses should be cunningly disposed along the streets, neither too open nor too private; and there should be substantial common recreation grounds. It is hardly possible to imagine a prescription less like the one that produced the tenement houses pictured in figure 220.

The Riverside Improvement Company accepted the recommendations of Olmsted and Vaux, and work began in 1868. It was interrupted by financial

236. Aerial photograph of Riverside, 1967
(University Library, Champaign-Urbana)

problems and by the fire of 1871, but in the end a
little town emerged, looking much as Olmsted intended. As may be seen from comparing figures 235
and 236, Riverside survives today as a striking
anomaly in the general grid. The wooded areas have
been largely preserved and even extended, and very
few of the streets have been changed. On the ground,
it is possible to see that many of the original houses
also survive. The circle at the center of an irregular
cross to the west of the river is a later addition,
Brookfield Zoo.

Figure 237 shows the town's relationship to Chicago. It was one of the towns that sprang up along
the line of the Chicago, Burlington, and Quincy
Railroad (now Burlington Northern), which has a
station in the heart of Riverside. Most of the towns
in figure 237 are platted as quadrilaterals, but Clar-
endon Hills has a design reflecting the influence of
Olmsted and Vaux.

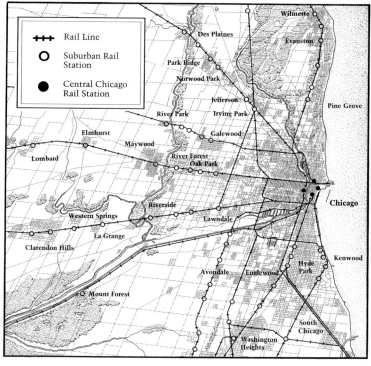

237. Map showing some Chicago
suburbs about 1876

Changes in the Chicago and Calumet River Systems

Figure 238 shows the mouth of the Chicago River as it is today, with much of the adjacent land available for development. The long creamy white building on the north bank is the North Pier Terminal, long more or less derelict but now renovated as a center for shops and restaurants. In the foreground is the former coastguard station, on the breakwater that controls access to the river mouth through a twentieth-century lock. This lock prevents Lake Michigan from flowing *into* the Chicago River, for in the nineteenth century the direction of the river was reversed.

Even with the crib system in place, very heavy rains could contaminate Chicago's water supply. One way to avoid this was to prevent as far as possible the flow of river water into the lake, by deepening the Illinois and Michigan Canal and so ensuring a brisk flow southwards, eventually into the Mississippi River. Work on deepening the canal began in 1866 and was completed in July 1871, just before the great fire. The effect was almost instantaneous; as one contemporary wrote, "Then there was the river—the horrible, black, stinking river of a few weeks ago, which has since become clear enough for fish to live in, by reason of the deepening of the canal, which draws to the Mississippi a perpetual flow of pure water from Lake Michigan."

Of course, the inhabitants of Chicago did not reflect, then or at any other time, that they were in fact flushing their sewage down to the towns on the canal and the Illinois River. The deepening of the Illinois and Michigan Canal and the reversal of the flow of the Chicago River greatly relieved Chicago's problems with sewage and fresh water. As the city inexorably grew, however, there were further incidents of contamination, when during heavy rains sewage flowed into the lake. Between 1889 and 1900 the city therefore undertook new and massive works to connect the Chicago River with the Illinois River (figure 239). When complete, the Sanitary and Ship Canal had required more land excavation than

238. Aerial photograph of the mouth of the Chicago River, October 1985

the building of the Panama Canal, and it remains the main water link between the Great Lakes and the Mississippi River.

Meanwhile the development of the industrial

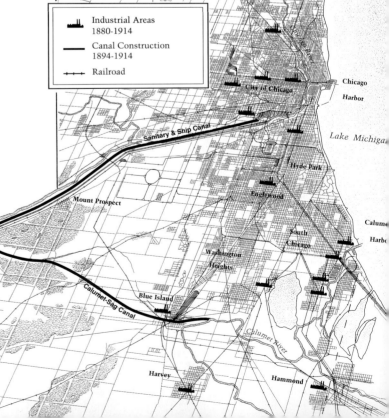

239. Map showing Chicago waterways and the Calumet industrial district, about 1900

area at the mouth of the Calumet River was going on apace. As figure 240 shows, the factories installed there were on a scale altogether larger than was possible in the cramped confines of the city, and eventually there was a great concentration of heavy industry, particularly along the Calumet River between Lake Michigan and Lake Calumet. This concentration of steel mills and other industrial plants led to increasing contamination of Lake Michigan and the Calumet River. In 1922 the Calumet Sag Channel was dug, connecting the Calumet area with the Sanitary and Ship Canal about twenty miles southwest of Chicago (figure 239).

The Chicago area now has a great variety of port installations, based on a triangular pattern. Barges come up the canal (on the Illinois Waterway) from the southwest, carrying bulk cargoes for Chicago's industries. Ships come down Lake Michigan, most often carrying iron ore for the smelters of the Calumet region. And ships sail out of the Calumet River, often carrying grain that has been stored in the elevators at Lake Calumet. The waterborne trade of Chicago is no longer very conspicuous to the casual visitor, but its tonnage remains impressive.

240. Aerial photograph of steelworks at the mouth of the Calumet River, 1970 (University Library, Champaign-Urbana)

173

241. Aerial photograph of Northwestern University, October 1985

A City of Universities

242. Chicago University (Otto Jevne and Peter Almini, *Chicago Illustrated*, Chicago, 1866–67; Newberry Library)

THE CHICAGO AREA IS EXTRAORDINARILY RICH in universities, both secular and religious. The earliest was Northwestern, founded in 1851 on a site by the lake, twelve miles north of the city. One of the university's founders was John Evans, and the town that grew up around the university was named Evanston in his honor. Figure 241 shows us the campus from the south, looking towards the Baha'i Temple and the long lake shore.

The campus has notable early buildings, like the brick block just north of the beach. But its most striking feature is the eighty acres taken from the lake in the 1960s. The landfill includes the large parking lot in the foreground and all the area to the north and east of it, including the promontory with the observatory. The enclosed lagoon is particularly successful in setting off the new buildings that line it to the west, behind the groves and clumps of willow that give this part of the campus a distinct flavor of the Backs at Cambridge University in England.

Five years after the foundation of Northwestern, another university was established by the lake, about three miles south of central Chicago. This institution, known first as Douglas University and then as Chicago University, seemed in the mid-

1860s, when our print was made (figure 242), to have a bright future. Alas, it had a heavy building debt and in 1885 went into a financial collapse from which it could not recover.

Six years later another university arose on the South Side, this time under the ample patronage of John D. Rockefeller. The 1890s were a great period of campus building in the United States, and the campus of the University of Chicago was one of the most remarkable. As may be seen from figure 243, which looks northwest, most of the early buildings on the north side of the Midway Plaisance were in the Gothic style, which gave a desirable unity and proved very flexible. They also have an undeniable aesthetic appeal, fulfilling the aim of one of the early professors who remarked, "The University takes pride in her laboratories, but she also covets for her students something of the charm of life in the cloisters of Oxford and Cambridge."

In fact, this "very latest English Gothic" came almost exclusively from buildings at Oxford. In 1900 the university's representative wrote to President Harper from there, remarking that he had "men working in Christ Church Hall taking measurements"; Hutchinson Commons, the Reynolds Club, and Mitchell Tower all came fairly directly from Oxford models, on which the other early buildings were also variations. Some people wondered if this enthusiasm for Gothic might not be going a little too far, and in 1901 Rockefeller asked, "Is it wise to use the cathedral architecture in a gymnasium?"

On the whole, this flexible and homogeneous style formed a remarkably harmonious set of campus buildings. The last great Gothic structure to be built was the Rockefeller Chapel, prominent in the middle of our view and dedicated in 1928. By then the "gray city," as somebody called it, had taken on the shape that it largely retains today.

243. Aerial photograph of the University of Chicago, October 1985

244. Drawing of the plan for the Chicago Academy of Sciences, nineteenth century (Chicago Academy of Sciences)

The Emergence of Libraries and Other Learned Institutions

245. View of the Chicago Public Library (*Chicago: The Great Central Market*, Chicago, 1921; Newberry Library)

CONTRARY TO POPULAR OPINION in the United States, Chicago was almost from the start a center for learned institutions. The Chicago Lyceum of 1834 was the first, followed by the Young Men's Association of 1841 and the Mechanics' Institute of 1842. The Chicago Historical Society was founded in 1856 and has continued on an ever-increasing scale. Figure 245 shows the Chicago Public Library (now the Cultural Center), which when it was built in 1897 was an outstanding expression of civic pride. Its superb marble staircases and elegant mosaics, the latter the work of Louis Tiffany, give a good idea of the civic aspirations of the turn of the century.

By then, too, sufficient capital was available to found many private institutions. The Chicago Academy of Sciences was a forerunner, in 1857; one of its founders was Robert Kennicott, whom we encountered as a child growing up at the Grove and as the botanist/zoologist in charge of collecting specimens along the track of the Illinois Central Railroad. The founding fathers of institutions like the Academy of Sciences had large visions, as may be seen from figure 244, although only the building at the left end was actually constructed. The academy has over the years been a tremendous resource for research and teaching about the flora and fauna of Illinois.

Specialized libraries also began to emerge. Many fine private collections were destroyed in the great fire, but others soon took their place. The Art Institute concentrated on books about the visual arts, and the John Crerar Library was founded in 1894 with a specialty in the natural sciences. In 1887 the Newberry Library was founded and at first collected very widely. Soon, however, the various librarians realized that they ought if possible to avoid duplication of effort, and a remarkable agreement was reached: the Crerar Library continued to concentrate on the natural sciences, while the Newberry collected principally works concerning history and literature; the libraries actually exchanged books, so as to rationalize their holdings.

It was of course difficult for privately endowed libraries like these to survive, and in the end the Crerar was transferred to the care of the University of Chicago. The Newberry has retained its indepen-

dence and serves a national, even international readership from its great grey fortress on Walton Street (figure 246). This building, in a style sometimes described as Spanish Romanesque, is like the Academy of Sciences only a portion of what was once planned, as may be seen from unfinished fragments of stone on its north side. It was originally going to cover the whole block, like a palace in Florence, but extends only along the south side of the block, facing Washington Square. It is an impressive building, one of many cases in which Chicago's fortunes were rapidly translated into cultural institutions.

246. Aerial photograph of the Newberry Library, May 1989

247. Aerial photograph of Saint Mary of the Lake, Mundelein, October 1985

Churches Ancient and Modern

AS FIGURE 220 SHOWS, the nineteenth-century churches of Chicago were among the dominant features of its skyline, and this visual dominance reflected the centrality of religion in the lives of many immigrants. Among the many varieties of Christians, Catholics from Germany, Ireland, Italy, and Poland were the most numerous. About 1860 the church with the largest congregation in Chicago was Holy Family, a long stone's throw from where Jane Addams would found Hull-House. Figure 248 shows Holy Family Church, with its "heavy Gothic" style. The parish was the early center of Jesuit activity, from which Loyola University and other institutions eventually developed. As a result of population shifts, the congregation of Holy Family has greatly declined, but it looks as if the great church itself will be maintained by special efforts.

Perhaps the most spectacular of Catholic leaders in Chicago was Cardinal Mundelein, who was appointed in 1915 and died in 1939. His reign—and

248. Aerial photograph of Holy Family Church, about 1970 (Seinwill Collection, Newberry Library)

249. Aerial photograph of the Baha'i House of Worship, Wilmette, October 1988

such a triumphalist term is not inappropriate—saw an extraordinary proliferation of ventures and institutions, of which the most remarkable was probably the seminary that he established at Saint Mary of the Lake (figure 247). There the children of immigrants would learn to be at the same time good "Americans" and good "Romans," and the architecture reflected this ambition.

The campus was built mostly in a colonial style, with the cardinal's residence resembling Washington's Mount Vernon, while the main chapel was modelled after a Congregational church in Connecticut. Many of the interiors, however, were imitations of Roman examples, reflecting the cardinal's aim to fuse the academic and disciplinary norms of Roman seminaries with a North American life-style. The seminary, now a university, remains a most striking

site, positively ducal in the scale and beauty of its grounds.

Another extraordinary religious complex is the Baha'i House of Worship at Wilmette, just above the harbor (figure 249). First planned in 1903, this spectacular temple was finished in 1953 under the direction of the architect Louis Bourgeois; the house in which he lived while directing the work stands across the road towards the lake from the temple. The building itself, set in grounds laid out with mathematical precision, consists of concrete panels mounted on a steel framework. The founder of the Baha'i faith was a Persian, Bahaullah, and the temple's style reflects this Persian influence. It is equally impressive seen from a distance, with its noble proportions, and seen close up, when the excellence of the cementwork is most striking.

250. Contemporary drawing of the exposition site (*The Book of the Fair,*
New York, 1894; Newberry Library)

The Columbian Exposition of 1893

IN 1851 THE CRYSTAL PALACE EXHIBITION held in
London began the fashion of international exhibitions, for which the world's cities contended.
The competition for the 1892 exhibition, to mark
the four-hundredth anniversary of the landing of Columbus in America, was stiff, but Chicago won it.
The city's representatives, who were leading businessmen, formed the World's Columbian Exposition
Corporation and made two key choices: Daniel
Burnham's firm for consulting architect, and Frederick Olmsted's for landscape architect.

A site was found by the infant University of Chicago, and work began. As figure 250 shows, a great
deal of earth was moved to form hillocks and lakes,
and the site was landscaped with over a million trees
and plants. More nations were induced to exhibit
than had ever before attended such a gathering, and
their displays were housed in neoclassical pavilions
colored—at Burnham's insistence—a uniform shade
of off-white. These buildings, which had so permanent and regal an air, were mostly constructed of
plaster of Paris on wooden frames (a building material known as *staff*); only the Fine Arts Building was
of brick and iron.

When it was finished in 1893, the White City
made a stupendous effect. Its buildings were not ar-
chitecturally innovative—visitors could go to downtown Chicago to see the latest technical advances—
but they showed how powerful a *planned* "city"
could be. In this sense, the exhibition seems to have

been the inspiration for Burnham's Chicago plan, which still affects the appearance of the city.

One of the immediate influences was the Intra-mural Railway, which probably inspired the construction of the Loop railway, built in 1897. The exhibition also signalled the emergence of Chicago (and the Midwest) on the global scene, as visitors from all over the world witnessed the extraordinary economic and cultural vitality of the city by the lake.

After the exhibition was over, the staff buildings were something of an embarrassment. Burnham thought the best idea would be ceremonially to burn them, and by accident most of them did burn down in 1894, leaving only the Fine Arts Building. Over the years this became somewhat dilapidated, but in 1930 it was reconstructed to house the Museum of Science and Industry (figure 251). This proved a prodigiously successful venture, exceeded in annual visitor numbers only in recent years by Washington's Air and Space Museum. The building remains a graceful reminder of the White City (figure 252), anchoring the northern end of Jackson Park, which exactly covers the former exhibition site.

251. Museum of Science and Industry being renovated, about 1930 (Percy Sloan Collection, Newberry Library)

252. Aerial photograph of the Museum of Science and Industry, October 1985

The Burnham Plan: Promise and Progress

CIVIC LEADERS WERE LONG AWARE of the need for parks, as the growing city swallowed up the prairie. Lincoln Park was founded in 1864, and in the late 1860s a group put forward a scheme for "parks near the city's west border and several south of the border, all linked by grand avenues." Humboldt, Garfield, and Douglas parks were established in 1869 and followed by Jackson and Washington parks. But these green spaces, although welcome oases on the tightly packed checkerboard, did not form part of a general plan and were quite dwarfed by the scale of the intervening streets.

Following his triumph at the 1893 exposition, Burnham was persuaded by members of the Commercial Club to try his hand at a general plan for the city. By 1909 this plan was published, creating a great sensation. Even today, leafing through the reprint of

254. Drawing by Jules Guérin from the Burnham Plan (Newberry Library)

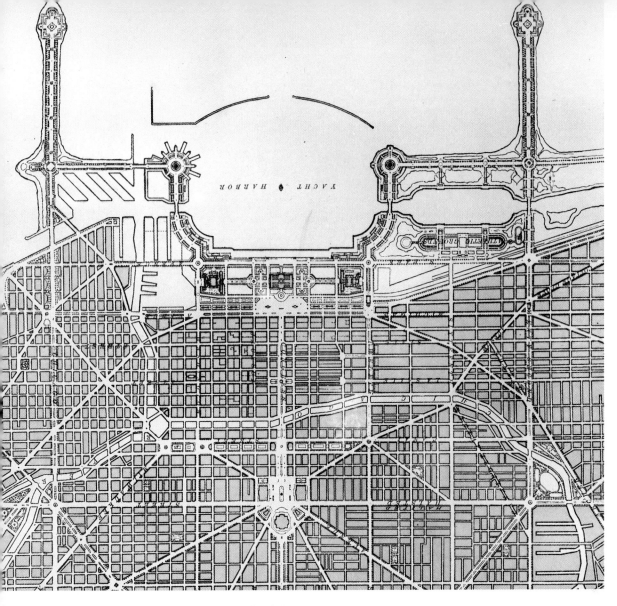

253. (above, left) Aerial photograph of the central area
of the city, about 1970 (Newberry Library)

255. (above) Map of the city from the Burnham Plan, 1907

the original, one can sense the immense amount of
work and thought that went into the plan, drawing
on the experience of all the major European cities.
The plan was also elegantly presented, using imaginary views by the watercolorist Jules Guérin (figure
254). Burnham wanted to break into the central grid
by organizing a system of radial and concentric bou-
levards (figure 255), rather as Haussmann had done
in Paris. This would have greatly eased the flow of
traffic, but proved very hard to carry out, so that in
the end only the Ogden Avenue extension was
undertaken.

At the lakefront Burnham envisaged a series of
parks on the mainland and on specially constructed
offshore islands (figure 254). The islands have not
been constructed, though in a sense the Northwestern landfill (figure 241) and the proposed landfill by

Loyola University are moves in that direction. But
the lakefront has indeed been preserved and laid out
in a long series of parks; the heart of this development is Grant Park, which emerged much as Burnham envisaged it (figures 253 and 255).

To enclose the central harbor area, he planned
two great piers, and the northern one was built as
Navy Pier (originally Outer Harbor Municipal Pier
No. 2) between 1914 and 1916. This pier has had a
checkered history. Between 1918 and 1930 it thrived
as a place for public events such as concerts and
dances. The Depression greatly reduced this activity,
and in 1941 the pier became a naval training station.
Between 1947 and 1965 it housed the Chicago
branch of the University of Illinois; since then this
major public amenity has been largely unused in
spite of various hopeful proposals.

256. Aerial photograph of the
Century of Progress Exposition,
1934 (Newberry Library)

257. Adler Planetarium under
construction, about 1929
(Percy Sloan Collection,
Newberry Library)

The Burnham Plan and the Century of Progress Exposition

ON THE SOUTH SIDE of the central harbor area, Burnham planned another great breakwater, and this was constructed in connection with the Century of Progress Exposition of 1934, celebrating one hundred years of Chicago's corporate existence. Figure 257 shows the area about 1929. Soldier Field is in place, as is the spanking new Field Museum. A wall extends into the lake to define the area of the new landfill, on which dredges can be seen at work.

Figure 256 shows the Century of Progress grounds from the northwest. They covered a large area on the mainland and most of the new landfill. On the northern end of the fill are two substantial new buildings, the Shedd Aquarium in the nearer circular area and the Adler Planetarium in the circular area sticking out into the lake. This northern promontory remains an attractive feature of the landfill, from which Chicago's skyline can be seen to great advantage (figure 258).

Burnham intended his offshore islands for parks and public recreation, and it seems a pity that the island south of the Adler did not revert to that use once the exhibition was over. Instead it has become Meigs Field, a small commuter airport. This is typical of the fate of much other lakefront property that might have become parkland. Soldier Field and McCormick Place have swallowed up much of the land in this area, and along the lakefront from Fifty-fifth Street in the south to Foster Avenue in the north, the multilane Lake Shore Drive eats into the area available for grass and trees. Considering the fate of most urban lakefronts, however, we ought perhaps to be grateful that so much remains for public use.

Down the years many have criticized the Burnham Plan. Lewis Mumford dismissed it as "municipal cosmetics," and soon after its publication a New York critic wrote that "Chicago will be tested far more by its housing than its lakefront. It won't do you any good to have that beautiful architectural scheme worked out for the lake shore if at the same time you have your slums and your bad tenement alleys, as they now exist on the west side." It is true that Burnham made little provision for low-cost housing, but that was, in a sense, outside the municipal limits of his plan. No doubt he expected that the federal, state, and municipal authorities would com-

bine to tackle the problem, as indeed they have done, with some degree of success. In the end, we ought to remember Burnham above all for offering a vision of the new city, even if it is only at the lakefront that this vision has been at all satisfactorily fulfilled.

258. Aerial photograph of the Adler Planetarium, October 1985

259. (opposite) Aerial photograph
of the city with aircraft
superimposed, about 1930
(Newberry Library)

260. Key to some buildings in
the aerial photograph of the city

1 Union Station
2 Chicago Daily
  News Building
3 Merchandise Mart
4 Board of Trade
  Building
5 Civic Opera Building
6 Palmolive Building
7 Wrigley Building
8 Tribune Tower
9 Buckingham Grain
  Elevator

Architectural Innovations in the Loop

IN THE PLATES ACCOMPANYING THE BURNHAM PLAN, the buildings are limited in height, giving a feeling rather like that of present-day central Washington or Paris. However, from the 1880s onwards Chicago architects were building daring new skyscrapers, and as time went by and downtown land values rose, so did the height of the buildings.

This photograph of about 1930, with the aircraft rather unconvincingly dubbed in, shows the city after the building boom that preceded the Depression. At the river mouth is a notable survivor, the grain elevator that we first encountered in 1857 (figure 179). Coming westwards upriver, we reach the new Michigan Avenue Bridge, flanked on its north side by the Wrigley Building (1924) and the Tribune Tower (1925). At the northern end of Michigan Avenue is the Palmolive Building, completed in 1930 and famous for the Lindbergh Beacon on the summit, which flashed as a guide to aircraft.

Away to the left, on the north bank of the river, the Merchandise Mart had just been completed. It was more brutal than elegant in style, and before the construction of the Pentagon was the world's largest building in floor area. At the bottom of this exceptionally clear photograph we see Union Station, whose immense waiting room was completed in 1925, and whose long train sheds stretch out north and south of the main building. The Burnham Plan has the apparent weakness of tolerating the continuation of seven downtown railroad stations, and this consolidation of some lines into Union Station was a promising development.

On the opposite bank of the river, the large white building is the Civic Opera House, funded by Samuel Insull and opened in 1929—just in time. That same year also saw the completion of the Board of Trade Building, capped by a statue of Ceres, Greek goddess of grain. This extraordinary burst of building in the 1920s, carrying on a process begun in the 1880s, quite transformed the appearance of the city. No longer could grain elevators and churches dominate the skyline; now, as at New York but nowhere else in the world, the skyscraper dominated the central urban scene.

261. An early image of Wacker Drive (*Chicago: The Great Central Market,* Chicago, 1921; Newberry Library)

The Coming of the Automobile City

THE CHAIRMAN OF THE CHICAGO PLAN COMMISSION, Charles Wacker, abstracted from the Burnham Plan a document called *Wacker's Manual of the Plan of Chicago,* which was used for many years in the eighth and ninth grades of Chicago's public schools. Thus at least during the first half of the century, many of Chicago's citizens knew about the main provisions of the Burnham Plan and could be expected to give their political support to proposals for implementing them.

The plan made provision at various points for a two-level pattern of streets, and Wacker's name is particularly associated with the two-level drive that was installed during the 1920s along the south bank of the Chicago River, where South Water Street had run (figure 261). The new Michigan Avenue Bridge was also bilevel, running into a system of underground roads beneath Michigan Avenue; these allow the motorist to cross Chicago quickly, for they are never as congested as the upper-level streets.

In 1937 the river mouth was spanned by the Outer Drive Bridge. This considerable accomplishment led to a remarkable example of faulty road engineering.

The Burnham Plan was very specific in affirming that "in laying out routes, no bad kinks or sharp turns should be tolerated," but the southern approach to the Outer Drive Bridge had to be designed with the famous S-bend, which plagued half a century of motorists. Figure 262 was photographed in the early afternoon, as we can tell from the shadows, and at that time a dense southward flow of vehicles was impeded at each of the bends. Today, after much landfill engineering, the S-bend has been converted into an elegant curve.

Figure 262 also shows us, to the west of the Outer Drive on the south bank of the river, the abandoned area where the Illinois Central tracks once fed the grain elevators; this strategic central area is still awaiting development.

Burnham hoped for a series of radial roads fanning out from the city center; though few were built, he would have approved of the great extension of Congress Street out to the west, where it becomes the Eisenhower Expressway (I-290). Figure 263 offers a striking view of this artery, which, as it reaches the city, dramatically passes under the main post office.

262. Aerial photograph of the Lake Shore Drive S-curve, about 1970 (Newberry Library)

263. Aerial photograph of the Eisenhower Expressway, looking east, about 1970 (Newberry Library)

II

THE STATE COMES OF AGE

264. Aerial photograph of the capitol in 1939 (Chicago Historical Society)

Introduction: The New Capitol

IT WAS ABOUT 1890 that Chicago became the Second City and Illinois became the third state, ranked by population. The state's economic resources were formidable, for in addition to a thriving agriculture and manufacturing centers, it added after the 1870s a huge coal-mining sector and soon developed the oilfields of the southeast.

To reflect this burgeoning demographic and economic power, the State Capitol of 1853, so imposing in its day (figure 83), was in 1888 replaced by a new and much larger building (figure 264). Its plan is like a Latin cross, with a central tower nearly four hundred feet high, which can be seen for miles around. The outer walls are made of limestone from Joliet and Lemont, and the interior is sumptuously marbled. In this aerial photograph, taken in 1939, the few cars are neatly parked along the edges of the roads, and the area to the west (top background) is still exclusively residential.

By the end of the nineteenth century, virtually all the potentially productive land in Illinois was under cultivation; farming techniques were constantly being improved and developed in the state's agricultural research centers. As figure 265 shows, this agricultural production is supplemented by a considerable number of manufacturing centers. An extensive mining sector produced coal in many locations throughout the state, and sand and gravel predominantly in the area formerly covered by the Wisconsin glacier.

This economic activity relied on an extensive shipping system, as the railroads came to be supplemented and often ousted by a vast network of roads able to carry the largest trucks. For bulk cargo there was some further development of the waterways, and for passenger and light freight a large system of transport by air.

Illinois has never been a state with a large number of federal defense establishments, but it does have, until 1990 at least, a few very large ones such as the federal arsenal at Rock Island, the Great Lakes Naval Training Base just south of Waukegan, and Chanute Air Force Base at Rantoul (figure 287). The U.S. Department of Energy also maintains the National Accelerator Laboratory, near Batavia, where fundamental research into the nature of matter is carried on.

The other great physics research institution in the state is the Argonne National Laboratory, administered by the Associated Midwest Universities; there the primary focus is on the peaceful uses of nuclear energy. The state has a considerable number of nuclear power plants, which pose the usual problems of balancing power needs against the degradation of the environment. As Illinois approaches the twenty-first century, the resolution of these problems becomes more and more pressing.

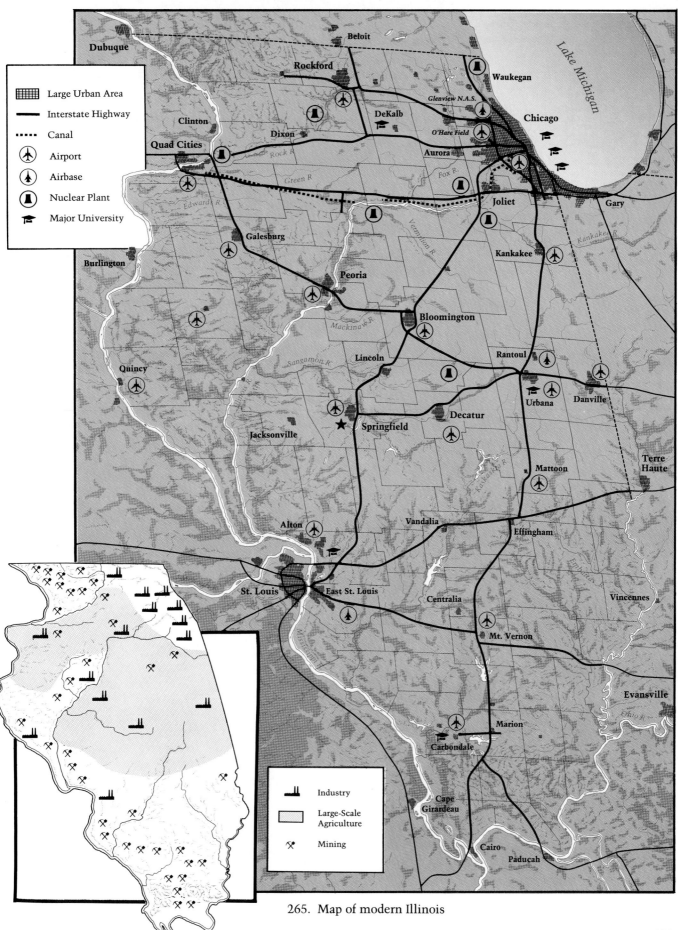

Legend (upper left):

- Large Urban Area
- Interstate Highway
- Canal
- ✈ Airport
- ✈ Airbase
- Nuclear Plant
- Major University

Legend (lower center):

- Industry
- Large-Scale Agriculture
- Mining

Dubuque
Beloit
Rockford
Waukegan
Clinton
Glenview N.A.S.
DeKalb
Chicago
Dixon
O'Hare Field
Aurora
Quad Cities
Rock R.
Green R.
Fox R.
Joliet
Gary
Edward R.
Kankakee R.
Galesburg
Kankakee
Burlington
Peoria
Vermilion R.
Mackinaw R.
Bloomington
Lincoln
Rantoul
Quincy
Sangamon R.
Urbana
Danville
Jacksonville
Springfield
Decatur
Terre Haute
Mattoon
Kaskaskia R.
Vandalia
Effingham
Alton
Centralia
Vincennes
St. Louis
East St. Louis
Evansville
Embarras R.
Mississippi R.
Mt. Vernon
Marion
Ohio R.
Carbondale
Cape Girardeau
Cairo
Paducah

Lake Michigan

265. Map of modern Illinois

The State's Industrial Centers

FROM THE START Illinois had the basic compo-
nents for developing a great manufacturing sector: easy access to raw materials, abundant fuel,
and good communications, at first by water. In the
middle of the nineteenth century several new elements combined to encourage the emergence of factories in the state. First, the population doubled between 1850 and 1860 and continued to grow after
that, providing not only abundant labor but also a
large market. The coming of the railroads, described
in chapter 8, linked Illinois not only with the established markets of the East, but also with the raw materials and emerging markets of the West. Finally,
the Civil War period saw a vast increase in demand
for military and other supplies, which Illinois was
well placed to satisfy; in this period the import of
foreign manufactured goods was restricted by war
conditions and by tariffs.

The result was that by 1880 the value of goods
manufactured in Illinois exceeded for the first time
that of agricultural produce. Perhaps as much as
three-quarters of this manufacturing capacity was
based in Chicago, but large manufacturing centers
also emerged elsewhere in the state, particularly at
Peoria, Joliet, East Saint Louis, Rockford, and the
present Quad Cities area (Rock Island and Moline).
Each of these centers became known for some particular activity (for example, Peoria was known first
for distilling and later for the Caterpillar Tractor
Company), but each also developed a wide range of
interdependent manufacturing ventures.

Figure 267 shows the John Deere Plow Company
factory at Moline about 1905. Its extent is impressive, as was its production, which at this period, as
the caption claims, reached "one complete imple-
ment for every working minute of the year." More-

266. (opposite) Works of the Joliet Iron and Steel Company (*Combination Atlas of Will County*, Elgin, 1873; Newberry Library)

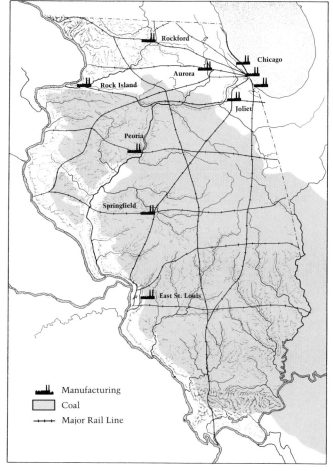

267. (above) Advertisement for the John Deere factory at Moline (*Atlas of Rock Island County*, Davenport, 1905; Newberry Library)

268. (left) Map showing major industrial areas of Illinois and coal region in the 1880s and 1890s

over, this was only one among several huge factories at Moline, including the Velie Carriage Company and the Moline Wagon Company.

Joliet remained a center for the mining of limestone, but its excellent transport facilities also allowed it to develop a wide range of manufacturing enterprises. One of the largest was the factory of the Joliet Iron and Steel Company, shown in a view of 1873 (figure 266). In the foreground, limestone is being unloaded from the barge on the Illinois and Michigan Canal. A train is pulling in from the right, laden with what looks like iron ore. On the ground, left of center, are stacks of the rails produced at this plant. The quantity of smoke is impressive, as is the scattered nature of the workers' houses in the background. Today this factory is being investigated by industrial archeologists, for possible development as a historical site.

269. Kempton (*Historical Atlas of Ford County*, Chicago, 1884; Newberry Library)

270. (left) Paxton (*Historical Atlas of Ford County*, Chicago, 1884; Newberry Library)

271. (right) *Business Property of A. C. Parker* (*Illustrated Atlas Map of Iroquois County*, Edwardsville, 1884; Newberry Library)

The Golden Age of the Small Towns of Illinois

THE GOLDEN AGE OF THE SMALL TOWN was the last quarter of the nineteenth century, a time when rural and metropolitan populations were much more in balance than today. The earlier struggle for existence, typified at New Salem, was over and the automobile had not yet ended rural isolation and accelerated the rush to the cities. During this brief period, small towns were self-sustaining places whose residents did not have to venture far from their front doors for most of the necessities of life. The towns were often beautiful and peaceful as well.

Kempton, in Ford County, is an example of the smallest type of town, having about one hundred inhabitants (figure 269). Basic to its existence was the railroad depot, with its attendant elevators and lumberyard, reminders of the town's economic foundation. The short line of shops no doubt included a tavern and a general store that provided just about everything the local farmer or town dweller needed, as well as a meeting place. A town usually had a church like the one visible at the end of the main street; this would be the other center for social life as well as a place for worship.

Kempton's school was located a short distance out of town, but a town of the next order of magnitude would have at least one school within its bounds, as well as a properly appointed post office and perhaps a small factory or two. It does not seem possible to say accurately at what stage of growth a town might attract a doctor, but it was probably rare to have one in towns of fewer than one thousand people.

Kempton has what looks like a boarding-house by the tracks; once the population grew above one thousand the town would have a proper hotel, a substantial row of shops like the ones shown in figure 271, and perhaps a courthouse. It might boast elegant homes like the ones in Paxton shown in figure 270. One of the delights of these small Illinois towns even now is their charming Victorian homes, many of which seem to have survived in their original shape.

If a town continued to grow, it might eventually attract public institutions such as a college and a hospital and would be likely to have its own printer and newspaper. At that size, towns benefited not merely from their convenience as market centers, but often from some kind of specialized manufacturing. In a sense specialized manufacturing was widespread, for local shops often made such foodstuffs as sausages and cheese or could produce clothing and shoes on the premises.

All this changed with the coming of the automobile, which made it possible to think of excursions far beyond the ten miles or so easily reached by the horses and buggies seen in figure 269. Many small towns dwindled almost to nothing, and even sizeable places of several thousand inhabitants found that many of their activities were relocated to larger urban centers. The golden age of small towns passed and shows no signs of returning.

272. Aerial photograph of the university campus,
September 1986

273. Key to the aerial
photograph of the campus

1 Huff Gym
2 Armory
3 Library
4 Morrow Plots
5 Altgeld Hall
6 Illini Union
7 Wright Street

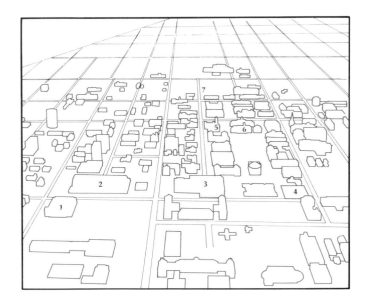

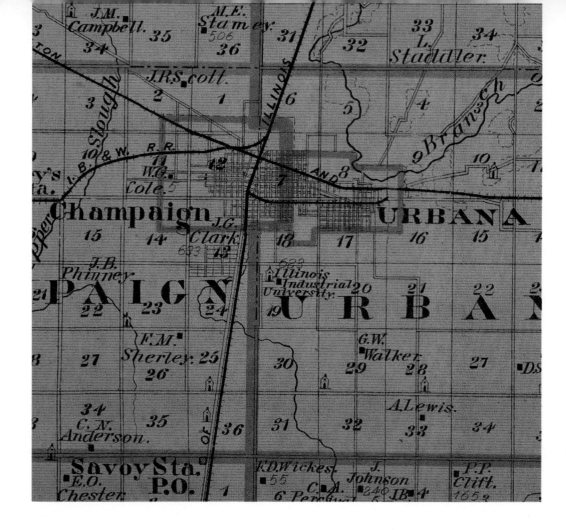

274. Detail of Champaign-Urbana (*Union Atlas*, 1876; Newberry Library)

The University of Illinois at Champaign-Urbana

URBANA WAS FIRST SETTLED IN 1822 and in 1833 became the county seat of Champaign County. It lay at the heart of an immensely productive agricultural region but in 1854 was bypassed when the Illinois Central Railroad was building its line southwards from Chicago. Figure 274 is a map dating from 1893, showing the relative positions of Urbana and the Illinois Central line (the east-west line was a later addition). The decision not to pass through the town meant that another center began developing around the depot, two miles west of Urbana; it eventually became known as Champaign.

The Land Grant Act of 1862, signed by President Lincoln, provided public land to support higher education, and the Illinois Industrial College was one of sixty-eight institutions that came from the act. Urbana was chosen for the site, and the college opened in 1868. Six years later University Hall was built on the site of the present Illini Union (figures 272 and 273), and in 1897 Altgeld Hall was completed, for use as the library. It is the only building on campus in the Romanesque style, and its red roof is prominent in the aerial photograph. It lies immediately to the east of Wright Street, which separates Urbana in the east from Champaign in the west (see the "Corporation Line" in figure 274).

In 1926 a new library was built, to house the university's remarkably extensive collections. It conformed to the modified Georgian style of most of the university's buildings around the main court area and has been extended several times to the west, as the aerial photograph shows. The Georgian style seems very appropriate for the university's massive buildings and gives the campus a unified feeling that it would otherwise lack.

No doubt the finest building in this style is the Illini Union, which replaced University Hall at the north end of the main court. Begun in 1938, it was completed in 1941 with help from the federal government; it remains the center of student life and offers accommodations for visiting parents and scholars. Although it has long since outgrown its "industrial" origins, the University of Illinois remains strong in the agricultural sciences. Symbolic of this long-standing specialty are the Morrow Plots, first planted in 1876 and retained at the heart of the central area.

The State Penitentiary at Joliet

STATES NEED PRISONS as well as universities, and by 1855 the state penitentiary at Alton was getting overcrowded. In 1857 the Illinois legislators commissioned the prominent Chicago architectural firm of Boyington and Wheelock to design a new penitentiary at Joliet. The architects chose a distinctive medieval style for the prison, as was then the fashion in England. It was completed in 1869 at the immense cost of $2 million.

Figure 276 shows the prison in 1873. In the foreground prisoners in their striped uniforms are planting trees, while three elegant gentlemen just beside them survey the scene. Here and there on the top of the walls are armed guards, while inside the prison a group of four convicts is being marched in lockstep. A carriage approaches the gate in the wall on the right, which is a sort of holding area that allows all traffic to be searched.

As figure 275 shows, the prison today has much of its original form. The main building seems to be quite unchanged, and even some of the internal structures seem to be the same. No doubt plenty of stone and plenty of labor is always available to keep the old buildings in repair and add new ones in an appropriate style.

275. (above) Aerial photograph of the state penitentiary in April 1987

276. (right) View of the state penitentiary (*Atlas of Will County*, Chicago, 1873; Newberry Library)

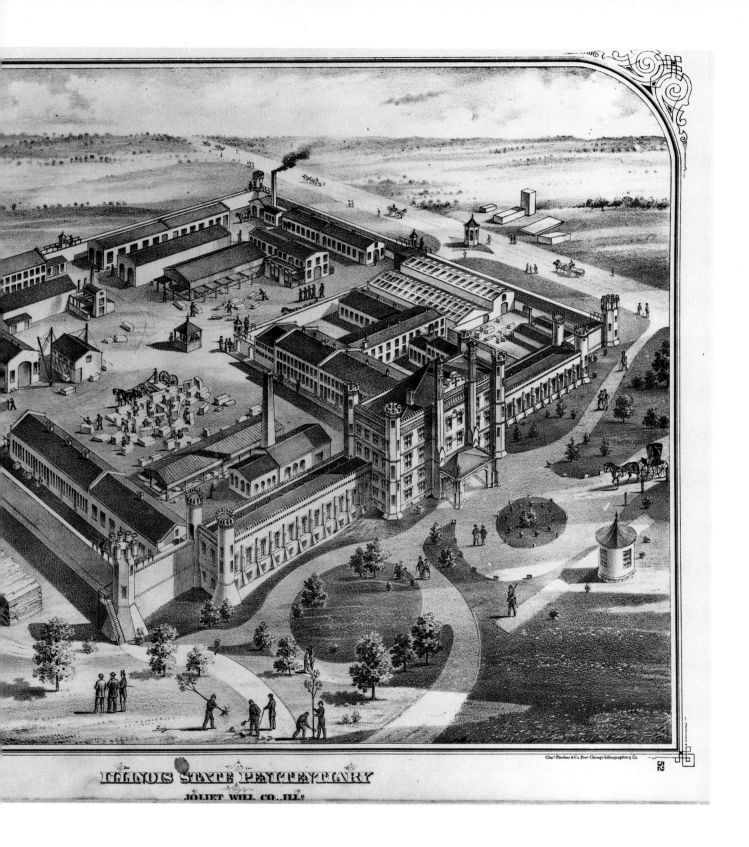

ILLINOIS STATE PENITENTIARY

JOLIET WILL CO., ILLs

Cha⁵ Shober & Co. Pres⁵ Chicago Lithographing Co.

52

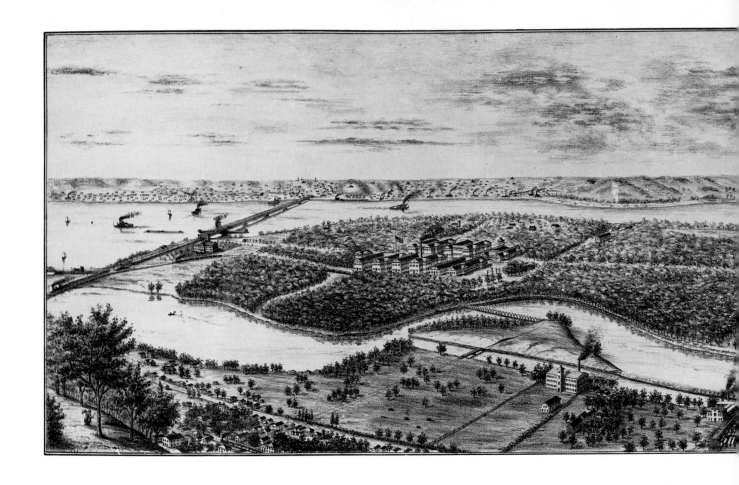

The Federal Arsenal at Rock Island

AS EARLY AS 1862 it was decided in Washington that Rock Island, the old site of Fort Armstrong, should be the site of a national arsenal. Work on its first building began during the Civil War, when part of the island became a prison for Confederate prisoners, some of whom are buried in the cemetery still to be seen there.

After the war, work began on the central range of buildings. In figure 277, from about 1876, just six of these "shops" were completed; four more were soon built to the right of the existing ones. In the aerial photograph the complete row may be seen. These handsome, massive Georgian buildings were faced with Joliet limestone. Over the years there have been additions to the central range, but we can easily identify the original buildings, which were U-shaped, with the bottom of the U backing onto the central roadway.

The arsenal had its own dam for generating electricity on an arm of the Mississippi River and produced enormous quantities of guns and other supplies for various wars, beginning with the Spanish-

American War of 1898 and continuing to the present day. Often the arsenal was a leader in military technology; during the period between the world wars, for instance, it led the way in research and development of prototype tanks. Today a museum on the site displays a great variety of weapons.

Most of the features marked on the nineteenth-century plans of the arsenal can still be seen, though there have been extensive additions, best seen in figure 89. The house of the commandant ("Quarters 1") is particularly conspicuous with its tower. In figure 278 it is at the top left, by the river, with a circular drive in front of it; the same house may just be seen in the trees behind the central range of buildings in figure 277. The arsenal was commanded over the years by many officers of distinction, including General Thomas Rodman, after whom the widely used Rodman guns were named. Rodman was probably the most important figure in the history of the arsenal, for he was largely responsible for the development of the nineteenth-century master plan; his tomb may be seen at the east end of the island.

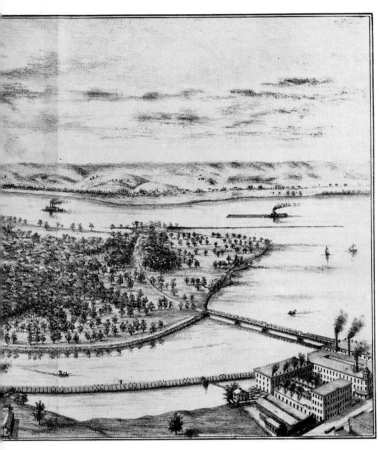

277. General view of the arsenal (D. W. Flagler, *A History of the Rock Island Arsenal*, Washington, 1877; Newberry Library)

278. Aerial photograph of the central buildings in the arsenal, October 1988

279. Aerial photograph of a lock on the Illinois Waterway, October 1988

280. Aerial photograph of an aqueduct on the Hennepin Canal, October 1988

Waterways of Illinois

THE ILLINOIS AND MICHIGAN CANAL, which was so important for the establishment of Chicago, carried a reasonable amount of freight from its opening in 1848 until the 1880s, when its annual tonnage declined. The section between Chicago and Lockport was in 1900 replaced by the much deeper Sanitary and Ship Canal, but that left the narrow old canal to the west, between Lockport and Utica.

In 1919 work therefore began on a western extension of the Sanitary and Ship Canal, and in 1933, with the help of the federal government, the work was finished. Many large locks were constructed by the U.S. Army Corps of Engineers in the course of all this work, and figure 279 shows one of them. We are looking southwards, with the Illinois River flowing from left to right. A tow of six barges is in the lock, which is full of water at the upper level. The river is running fast through the mill races at the left, separate from the canalized channel.

As figure 265 shows, there are two other substantial canals, the Calumet Sag Channel and the Hennepin Canal. The Calumet Sag Channel is much the more important, for since 1922 it has offered a connection between the Mississippi River and the great concentration of industry in the Calumet region. Barges coming up the Mississippi and Illinois rivers pass into the Sanitary and Ship Canal, then along the channel into the Calumet region. This passage, together with the branch that emerges at Chicago Harbor, is known as the Illinois Waterway. Along this waterway move huge quantities of bulk goods such as coal, grain, and petroleum products, on barges like the ones shown in figure 279. In fact, Illinois has twelve hundred miles of inland waterways, the longest network in the nation.

The Hennepin Canal is another story. It runs along the former bed of the Mississippi River, connecting the Illinois River at Hennepin with the Mississippi River near Rock Island. First mooted in the 1830s, it was eventually constructed by the U.S. Army Corps of Engineers between 1890 and 1907 but never carried any substantial traffic. The Corps of Engineers ceased to operate it in 1951, and since 1970 it has been under the jurisdiction of the Illinois Department of Conservation. Figure 280 shows the canal as it crosses a river on an aqueduct; it has become a valuable recreational resource, offering boating and fishing.

281. Aerial photograph of typical countryside in the Illinois corn belt, October 1985

The Agriculture of Modern Illinois

OVER 80 PERCENT OF THE LAND area in Illinois is farmed in one way or another, taking advantage of a climatic range running from 210 frost-free days per year in the south, to only 150 such days in the north. General agriculture (still including some sheep farming) takes place around the southern, western, and northern borders of the state, while the central prairie is the area of corn and soybean production, which most people think of as the typical Illinois countryside. The large fields of this region are conspicuous in figure 282.

From the air (figure 281) this central area is remarkably beautiful, with farms scattered widely on a landscape resembling a vast checkerboard, or perhaps a painting by Mondrian. The patterns of landholding conform basically to the township and section lines laid out by the early surveyors, but as time goes by the fields tend to get larger and larger, with the progressive mechanization of all agricultural processes. This tendency perhaps leads to more efficient production, but it may be at some cost in proper land conservation, and certainly is at the expense of the old-fashioned family farm.

Figure 283 shows us an area of this land in Champaign County, one of the most fertile regions of central Illinois. The photograph was made in 1982 for the Department of Agriculture (note this information on the lefthand edge), and includes the little town of Royal, through which the Missouri Pacific Railroad runs (though the alignment of the town makes no concession to this diagonal).

Note at the northern (top) end of the photograph a complete set of one-mile by one-mile sections, some of them divided into huge fields. This neat stacking of sections breaks down on the southern edge of this range, where the township-and-range system has an offset (see chapter 4 on this term). This makes the prominent north-south road on the eastern edge of the photograph take an uncharacteristic S-turn, and dislocates the neat section squares for much of the view. It would be interesting to examine a sequence of such photographs for the past fifty years, to assess the degree of consolidation among the fields and farms; it surely would be considerable.

282. (opposite) Detail from the LANDSAT image of the state, 1982 (Illinois State Geological Survey and Northern Illinois University)

283. (above) Aerial photograph of a typical field pattern in Champaign County, Illinois, September 1982 (University Library, Champaign-Urbana)

284. Aerial photograph of a typical road in the Illinois countryside, September 1986

Highways of Illinois

THE EARLIEST ROADS in Illinois usually followed Indian trails, often along ridges. Once the township-and-range system became established, the new roads usually ran along the section lines at mile intervals, in a north-south or east-west direction. Figure 284 shows a typical example of one of these roads, with farms spaced along it at roughly regular intervals. These roads were established during the nineteenth century, when the horse was the chief means of local transport for people and for goods.

The coming of the automobile did not materially alter this part of the road system, except that from 1914 onwards automobile-license fees paid for paving the roads. In 1914 there was less than one mile of concrete road in the whole of Cook County outside Chicago, but during the 1920s the whole state was covered with a network of paved highways (figure 285).

In 1956 the Federal Highway Administration began building the National System of Interstate and Defense Highways, in collaboration with the highway departments of the various states. These highways were constructed largely with federal money, but the states decided upon their alignment and are responsible for their upkeep. The Illinois Department of Transportation (IDOT) looks after eighteen hundred miles of interstate highway, as well as seventeen thousand miles of state highway. This is the third-largest transport network in the states, and in 1988 IDOT spent about $1 billion on its maintenance. It is heavily used by automobiles, and each year sustains a traffic of roughly ninety billion tons of freight.

The state also has a number of tollways; figure 286 is an aerial view of the East-West Tollway (I-88), looking east towards Chicago from a point roughly above the Fox River. Note the tollbooths at the bottom center of the photograph. A view like this gives some idea of the vast scale of these highway enterprises.

286. (opposite) Aerial photograph of the East-West Tollway, October 1988

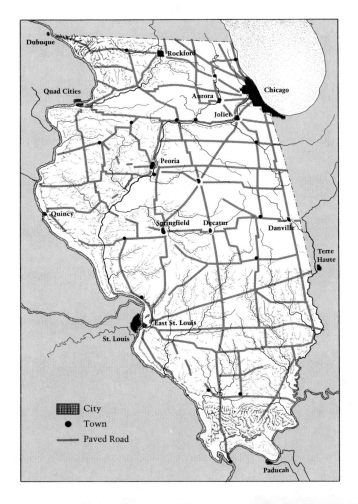

City
Town
Paved Road

285. Map showing the paved roads in Illinois, about 1925

287. Aerial photograph of the central area at Chanute Air Force Base, September 1986

The Aeronautical Revolution

As early as 1855 the balloon *Eclipse* was flown at Chicago, and in the following years balloons became commonplace at carnivals and exhibitions. Towards the end of the century they were joined by airships, or dirigibles, and by then pioneers like Octave Chanute were getting close to solving the problems of heavier-than-air flight.

After a career as a railroad engineer, Chanute moved to Chicago in 1889. He conducted many experiments with flying machines, and corresponded with the Wright brothers and with other pioneers like Lilienthal. After the Wright brothers' success at Kitty Hawk in 1903, exhibitions of powered flight became common, and several companies were formed in Chicago to build aircraft. In 1911 and 1912 the city hosted international aviation meetings, in which most of the leading flyers in the United States participated.

When the United States entered the First World War, it was necessary to train large numbers of pilots as quickly as possible. In Illinois flying schools were established at Rantoul (Chanute Field) and near Belleville (Scott Field). Chanute Field eventually became a great center for technical training, during and after the Second World War. Figure 287 shows its main hangars in the summer of 1986, when an interesting selection of aircraft was parked on the apron, probably for a visitors' day. At the time of writing, the future of Chanute as an air force base is in doubt.

The experience gained during the First World War made it possible by 1919 to inaugurate mail service by air, for which Chicago was a major staging area. In 1927 Midway Airport was dedicated, and it eventually became the world's busiest airport. During the 1920s and early 1930s it was not yet clear that the future lay with aircraft rather than with airships. In 1929 Chicago was visited by one of the most famous of the airships, the *Graf Zeppelin*, in the course of a world tour. Figure 288 shows it passing over the cribs in Lake Michigan. Four years later the Italian air minister Italo Balbo led another visit to Chicago, at the head of a squadron of twenty-four seaplanes which must have been an astonishing sight.

During the Second World War the Douglas Aircraft Company built great quantities of transport aircraft at Orchard Place, in the northwest suburbs of Chicago. In 1946 this land was transferred to the city, which was beginning to need a second airport to supplement Midway, heavily stretched with nine million passengers handled in 1955. This new airport was built at Orchard Place (hence the call letters ORD) and opened in 1959 as O'Hare Field. It has an unusual plan, easily seen in figure 289. Instead of the satellite terminals common elsewhere, its terminals all lead off a central core, accessible by automobile and rapid transit. A good deal of walking is inevitable at airports, but the central-core plan at O'Hare seems to minimize this movement, while leaving every part of the airport accessible by foot.

Because its cramped site cannot accommodate the longer runways needed by large passenger jets since 1959, Midway handles mostly short-haul routes, and O'Hare has taken over the claim to be the world's busiest airport. We should remember, however, that it is only one of over one hundred commercial airports in Illinois, a total exceeded only in California.

288. *Graf Zeppelin* on a visit to Chicago in 1929 (Newberry Library)

289. Aerial photograph of O'Hare Field, 1970 (Seinwill Collection, Newberry Library)

290. Aerial photograph of housing development outside Chicago, October 1985

291. Aerial photograph of highway advancing into the countryside about 1960 (Illinois Department of Transportation)

292. Aerial photograph of coal-mine tailings by the Illinois River, November 1988

The Dilemmas of Growth

ALTHOUGH SOME COUNTIES have experienced considerable losses in population, the state as a whole has continued to grow from census to census, particularly in the area around Chicago. This constant growth has led to ever-increasing pressures on the state's natural resources and to serious problems of pollution. In the industrial area to the south of Chicago in particular, many landfills contain substances harmful to the water supply. Often the contamination of streams and lakes can be detected and graphically demonstrated from the air, using infrared film, though we have no examples of that technique for this book.

As well as contaminating the land, air, and water, we have been using the arable land up at an alarming rate. Figure 290 shows a housing development outside Chicago, whose extended suburbs now reach almost solidly to the Fox River. Figure 291 shows the massive destruction caused by the interstate highways. In the foreground the advancing highway is about to annihilate yet another farmstead, while in the background the various access ramps and areas of dead ground show how vast is the acreage lost through this form of highway construction.

Illinois has also lost much land to mining, mostly of coal. Figure 292 shows an area by the Illinois River where tailings from a coal mine have kept an

293. Aerial photograph of the Ottawa "effigi tumuli," April 1986

area sterile and useless for decades. On the other hand, figure 293 shows a constructive use for mined-out land. Here a series of "effigi tumuli" have been carved by the earth sculptor Michael Heizer on land reclaimed and grassed by the Illinois Abandoned Mined Lands Reclamation Council. The council has reclaimed about one hundred mine sites in the state, but much remains to be done.

294. Aerial photograph of the main building at Fermilab, October 1985

The Atomic Era

ILLINOIS'S IMMENSE RESERVES of coal are mostly unsuitable for generating electrical power, since they are high in sulphur and consequently contribute to the problem of acid rain. Therefore, to meet the huge energy needs of the Chicago area, the Commonwealth Edison Company has built about a dozen nuclear reactors, on six sites.

The Dresden Nuclear Power Station was the nation's first privately owned nuclear power plant and was ready in 1960. It uses boiling-water reactors of relatively primitive design. The two units at Byron, shown in figure 296, are of the more advanced pressurized-water type, and each generates 1,120 MW. In some European countries huge buildings like these offensively dominate the landscape. But in the open space of Illinois they seem relatively inconspicuous, except that the steam from their cooling towers makes a very good landmark.

It was at the University of Chicago that the Italian nuclear physicist Enrico Fermi developed the first nuclear chain reaction. His name is now attached to the National Accelerator Laboratory near Batavia. This remarkable facility, run by a fifty-four-member university consortium for the U.S. Department of Energy, has a huge central building (figure 294), where its many research activities are administered. The heart of Fermilab, as the site is called, is the four-mile-round particle accelerator, whose outline in the earth may be seen even on satellite images (figure 295). This accelerator permits Fermilab's physicists, who come from many different countries, to conduct experiments aimed at understanding the nature of matter.

Much of the Fermilab complex is open to visitors, who can see presentations about the latest work and visit displays of Indian artifacts, gathered by some of the farmers who formerly occupied the site. When the federal government took it over, the farmhouses were brought together in an unusual village for researchers. Other attractions include a display of old farm implements and a bison herd; at Fermilab, we can meet up with Illinois past, present, and future.

295. (left) Detail from a LANDSAT image showing the accelerator ring at Fermilab, September–October 1982 (Illinois State Geological Survey and Northern Illinois University)

296. (below) Aerial photograph of the nuclear power station at Byron, April 1988

297. Aerial photograph of the capitol, September 1986

CONCLUSION

IN WRITING THIS BOOK we have tried to look at the history of Illinois literally from a new perspective, and in so doing are more convinced than ever of the advantages of the aerial view over many earthbound images. These advantages fall under three sometimes overlapping headings. First, aerial photographs at times allow us to detect sites of buildings or formations of which little or nothing remains; this is the case with figures 53 (Fort de Chartres), 196 (Belt Railway classification yard), and 198 (abandoned track). A satellite image like figure 1 allows us to see the geological bones of the state.

Figure 1 overlaps into our second category of advantages for the aerial view: in its extent it can reveal patterns that are not evident to the earthbound. The classic example of this phenomenon is the way in which Roman roads may be followed for great distances through the European countryside, even when from the ground they seem to have disappeared. In Illinois we have a striking aerial image of the National Road (figure 74) and the surviving French long-lots (figures 44 and 58), as well as the better-known example of the pattern imposed by the township-and-range system (figures 64 and 283). The great range offered by aerial photography can also be used to good effect in studying individual buildings; the Channahon locks (figure 137), the Grove (figure 145), the Graue Mill (figure 154), and the Cantigny mansion (figure 219), all take on a different significance when we can see them in their extended environment.

Finally, the aerial photograph can give us instant, accurate information about transient phenomena. Here its classic use is in the detection of pollution at sea, but it can also be used as in figure 13 to accurately count deer in a forest, or as in figure 257 to catch the Adler promontory at a precise stage in its construction, or as in figure 259 to show most of the city at a special stage in its architectural develop-

ment. Virtually all of these examples not only enable us to put the present into a wider context but also help us to see how tightly our present landscape and life-style are determined by the largely unseen patterns of the past, whether they are Indian trails in the Chicago street pattern (figure 24), or French long-lots, or township-and-range-determined field boundaries and roads.

Sometimes the raw material is hard to interpret, and we have added explanatory sketches (figures 260 and 273); in other cases no aerial view was available, and we have had recourse to imaginary aerial perspectives. These have the advantage of being much easier to interpret, in their three-dimensional quality, than a maplike or vertical photographic image, and of encouraging us to ask further questions as we drew them; they are a sort of heuristic device, leading on from one question to another.

In writing this book, we have been constantly aware of the tensions in Illinois between "Chicago" and "downstate." Sometimes, indeed, it has almost felt as if Chicago were an entity apart from the state, a sort of federal district on the shores of Lake Michigan. Certainly many of the inhabitants of each area have a curious mistrust and ignorance of the other, Chicagoans tending to think of downstaters as remote bumpkins, and downstaters often regarding Chicagoans as hopelessly corrupted by city life and urban pressures.

While both these views have a little truth, we hope in this book to have reminded Chicagoans that the whole early development of the state took place almost as far from Lake Michigan as it is possible to get, and to have shown downstaters that under the parking lots and high rises much of extraordinary historical interest survives at Chicago. We also hope to have demonstrated how closely the historical fortunes of each group were tied up with the development of the other.

The great point of contact between the two groups is Springfield, with its labyrinthine political processes. There the interests of the metropolis and of the farmland meet and clash and trade off in ways that are not always beneficial for either, though in a world where investment from overseas may be vital to economic survival, parochialism is more than ever out of place.

Our final figure shows the capitol at Springfield late on a fine evening in September 1986. Comparing it with the view of the capitol in 1939 (figure 264), we are reminded in a gentle way of what surely must be the greatest problem facing both city and state: the increasing pressure on natural resources.

Whereas in 1939 houses and gardens lay almost in the shade of the capitol's dome, by 1986 almost all the surrounding area was covered with concrete, for buildings and parking lots; indeed, one lot reaches up to the steps of the capital itself.

The photograph thus stands as a symbol of one of the state's most pressing problems. But it also stands as a symbol and measure of the state's progress, when we compare the capitol of 1853 (top right here and figure 83) with that of 1888. These have been years of extraordinary growth, and if they have brought many problems, they have also brought a remarkable improvement in the quality of life of many of the state's citizens.

APPENDIX

Sources and Methods for the Aerial Perspective Maps and Views

THE AERIAL PERSPECTIVE MAPS (figures 2, 3, 5, 15, 21, 25, 28, 35, and so forth) were made by drawing perspective grids and then transposing geographic information onto them from modern U.S. Geological Survey maps. Once these perspective maps were made, individual historic maps could be constructed by adding information from old county atlases, survey maps, vegetation maps, historical atlases, and other contemporary maps. This appendix will list such material used for the main perspective maps.

The aerial perspective views (figures 50, 52, 72, 114, 118, 141, and so forth) are in each case based on contemporary maps. For figure 50, for instance, the base map was Thomas Hutchins's map of 1766, and for figure 72 it was a plat map from *Lincoln's New Salem* by Benjamin P. Thomas (Springfield, Ill., 1934). Information from these base maps was then plotted onto a perspective grid, after which additional information was incorporated, using county atlas maps, other contemporary maps, satellite images, and aerial photographs. The next step was to draw buildings and structures in the perspective view, using illustrations of the actual towns and images of representative buildings from the same region. We visited many of the areas and took photographs of any extant structures.

Some of the other drawings, such as the Indian encampment (figure 20) and the steamboat (figure 80), were based on modern models, while others were derived from contemporary artifacts and full-sized reproductions. Some drawings needed a combination of sources; the image of the schooner in figure 117, for instance, is based on models, original builders' plans, old photographs, and contemporary accounts. Below we list the sources for the main views and drawings as well as for the maps.

Figures 15, 35, 62, 86, 144, and 265 (introductory maps of the state). Modern U.S. Geological Survey

maps of Illinois at 1:250,000. Nineteenth-century General Land Office survey maps. John M. Peck, John Messinger, and A. J. Mathewson, *New Sectional Map of the State of Illinois*, New York, 1851. Union Atlas Company, *Atlas of the State of Illinois*, Chicago, 1876. *Rand, McNally Commercial Atlas and Marketing Guide*, Chicago, 1900 and 1956 editions. Rand McNally and Company, *Illinois: Guide and Gazetteer*, Chicago, 1968.

Figure 18 (aerial perspective of the Cahokia Mounds site). Photographs from the Cahokia Mounds State Park and Museum. George E. Stuart, "Who Were the 'Mound Builders'?" *National Geographic*, 142 (1972): 783–802. *Handbook of North American Indians*, vol. 15, ed. Bruce Trigger, Washington, 1978.

Figures 25, 129, and 185 (maps of Chicago and the Illinois River Valley). Modern U.S. Geological Survey maps of Illinois at 1:250,000. Nineteenth-century General Land Office survey maps. James H. Rees, *Map of the Counties of Cook and DuPage*, Chicago, 1851. Union Atlas Company, *Atlas of the State of Illinois*, Chicago, 1876. Milo P. Quaife, *Chicago's Highways Old and New*, Chicago, 1923. Robert Knight and Lucius Zeuch, *The Location of the Chicago Portage Route of the Seventeenth Century*, Chicago, 1928. Irving Cutler, *Chicago: Metropolis of the Mid-continent*, Dubuque, Iowa, 1976. Harold M. Mayer and Richard C. Wade, *Chicago: Growth of a Metropolis*, Chicago, 1973. Michael P. Conzen and Kay J. Carr, *The Illinois and Michigan Canal National Heritage Corridor: A Guide to Its History and Sources*, DeKalb, Ill., 1988.

Figure 50 (aerial perspective of the village at Cahokia). Map of Cahokia about 1766, reproduced in Charles E. Peterson, "Notes on Old Cahokia," *Journal of the Illinois State Historical Society* 42 (1949): 193–208. Description by Philip Pittman in *The Present State . . .* , ed. J. F. McDermott, London, 1770;

Memphis, 1976. Maps in *Indian Villages of the Illinois Country*, ed. Sara Jones Tucker, Springfield, 1942. *Standard Atlas of St. Clair County, Illinois*, Chicago, 1901. *The French in the Mississippi Valley*, ed. J. F. McDermott, Urbana, 1965. F. Terry Norris, "Old Cahokia: An Eighteenth-Century Archaeological Site-Model," *Newsletter of the Center for French Colonial Studies*, 2 (1984). Photographs taken for this book at Cahokia and at Sainte Genevieve, Missouri.

Figure 72 (reconstructed view of New Salem). Plat map from Benjamin P. Thomas, *Lincoln's New Salem*, Springfield, Ill., 1934. *Illustrated Atlas Map of Menard County, Illinois*, n.p., 1874. John Mack Faragher, *Sugar Creek: Life on the Illinois Prairie*, New Haven, 1986. On-site photography at New Salem, and aerial photographs provided by Maynard Crossland of the Illinois State Historical Library.

Figure 114 (reconstructed view of Chicago in 1829). F. Harrison and W. B. Guion, *Map No. 1 of the Survey of the Michigan-Illinois Canal*, 1828. W. Howard and F. Harrison, *Map of the Mouth of the Chicago River Illinois*, 1830. "Map of Chicago in 1830" in Alfred T. Andreas, *History of Chicago*, Chicago, 1884. Chicago Historical Society, *Pictorial History of Chicago*, Chicago, 1930. "Chicago with the School Section" (1834) in Harold M. Mayer and Richard C. Wade, *Chicago: Growth of a Metropolis*, Chicago, 1973. John W. Larson, *Those Army Engineers: A History of the Chicago District, U.S. Army Corps of Engineers*, Washington, 1979.

Figure 118 (reconstructed view of the mouth of the Chicago River in 1839). "Plan of Chicago Harbor" (1839) in John W. Larson, *Those Army Engineers*, Washington, 1979. J. D. Craham, *Chicago Harbor and Bar*, Washington, 1858. Bird's-eye view of Chicago in 1857 by I. T. Palmatary, Chicago, 1857. Alfred T. Andreas, *History of Chicago*, Chicago, 1884. Paul Gilbert and Charles Lee Bryson, *Chicago and Its Makers*, Chicago, 1929. Herman Kogan and Lloyd Wendt, *Chicago: A Pictorial History*, New York, 1958.

Figure 141 (reconstructed view of the Illinois and Michigan Canal and Chicago River about 1858). James T. Baker, *Map of Chicago and Environs Exhibiting all Sub-divisions and Additions*, New York, 1854. W. L. Flower, *Map of Cook County, Illinois*, Chicago, 1862. Rufus Blanchard and Cram, *Guide Map of Chicago*, Chicago, 1868. Bird's-eye view of Chicago in 1857 by I. T. Palmatary, Chicago, 1857. Herman Kogan and Lloyd Wendt, *Chicago: A Pictorial History*, New York, 1958. Harold M. Mayer and Richard C. Wade, *Chicago: Growth of a Metropolis*, Chicago, 1973.

Figure 222 (aerial perspective of Block 15). *Tenement Conditions in Chicago: Report by the Investigating Committee of the City Homes Association*, ed. Robert Hunter, Chicago, 1901. Edith Abbott, *The Tenements of Chicago*, Chicago, 1936. Peter d'A. Jones and Melvin G. Holli, *Ethnic Chicago*, Grand Rapids, Michigan, 1981. Fire insurance atlases preserved at the Chicago Historical Society.

BIBLIOGRAPHY

General Works Useful in Studying the History of Illinois,
with a Selection of Books for Each Chapter

GENERAL WORKS

Angle, Paul, ed. *Prairie State.* Chicago, 1968. A very useful collection of impressions of Illinois.

Bach, Ira, and Susan Wolfson. *A Guide to Chicago's Historic Suburbs.* Chicago, 1981. A remarkable work of compilation covering a great number of interesting sites.

Bridges, Roger, and Rodney Davis. *Illinois: Its History and Legacy.* Saint Louis, 1984. Useful essays by experts on different aspects of Illinois history.

Carpenter, Allan. *Illinois, Land of Lincoln.* Chicago, 1968. Theoretically written for children, this book contains much novel visual material.

Chicago History. The periodical of the Chicago Historical Society.

Historic Illinois. The journal of the Illinois Historic Preservation Agency.

Howard, Robert P. *Illinois: A History of the Prairie State.* Grand Rapids, Mich., 1972. A sound and reasonable general history of the state.

Illinois MapNotes. Periodically published by the Illinois State Geological Survey. Contains much information about present and past mapping of the state.

Journal of the Illinois State Historical Society. The major journal devoted to all aspects of Illinois history.

Mayer, Harold M., and Richard C. Wade. *Chicago: Growth of a Metropolis.* Chicago, 1973. A richly illustrated account of Chicago's development.

Nelson, Ronald E., ed. *Illinois: Land and Life in the Prairie State.* Dubuque, Iowa, 1978. Contributions by experts to a geographical analysis of the state.

Rand McNally and Company. *Illinois: Guide and Gazetteer.* Chicago, 1969. Still the only recent general guide to the state's cities and towns: it derives from the WPA publication of the 1930s.

Sutton, Robert M., ed. *The Prairie State: A Documentary History of Illinois.* 2 vols. Grand Rapids, Mich., 1976. A rich collection of documents bearing on the history of Illinois.

———. *The Heartland: Pages from Illinois History.* Lake Forest, Ill., 1982. This paperback contains lively accounts of different aspects of Illinois history.

CHAPTER 1

Howard, Robert P. *Illinois: A History of the Prairie State.* Grand Rapids, Mich., 1972. Good chapter called "Prairie, portages, and Indians."

Nelson, Ronald E., ed. *Illinois: Land and Life in the Prairie State.* Dubuque, 1978. Extensive chapter on the physical environment, with full references to the literature.

CHAPTER 2

Bennett, John W. *Archaeological Explorations in Jo Daviess County, Illinois.* Chicago, 1945.

Cahokia Mounds Museum Society. *The Cahokia Mounds: A Guidebook.* Collinsville, Ill., n.d.

Carter, C. E. *Great Britain and the Illinois Country 1763–1774.* Washington, 1910. Describes the pressure in Great Britain and in the colonies for expansion across the Alleghenies, and the British government's successful resistance to it.

Clifton, James. "Billy Caldwell's Exile in Early Chicago." *Chicago History* (winter 1977): 218–28.

———. *The Prairie People.* Lawrence, Kansas, 1977. Survey of the Potawatomi.

De Kalb County Atlas. Chicago, 1905. Contains a good account of the life of Chief Shabbona.

Handbook of North American Indians. Vol. 15. Ed.

Bruce Trigger. Washington, 1978. Many useful chapters, especially "Late Prehistory of the Illinois Area," by Melvin Fowler and Robert Hall; "History of the Illinois Area," by Joseph Bauxar; "Illinois," by Charles Callender; and "Potawatomi" by James Clifton.

Hauser, Raymond. "The Illinois Indian Tribe: From Autonomy and Self-sufficiency to Dependency and Depopulation." *Journal of the Illinois State Historical Society* 69 (1974): 127–38.

Madden, Betty I. *Art, Crafts, and Architecture in Early Illinois.* Urbana, Ill., 1974. Good chapter on art before written history, with excellent section on the Piasa figure.

Matson, N. *Memories of Shaubena.* Chicago, 1882. Matson knew the Indian chief Shabbona and here tells what he learned from him.

———. *Pioneers of Illinois.* Chicago, 1882. Much interesting material, often gathered from people who knew the state before the great European influx.

Quaife, Milo P. *Chicago's Highways Old and New.* Chicago, 1923. A masterly summary by a great expert, with much material on the earliest period of European penetration.

Stuart, George E. "Who Were the 'Mound Builders'?" *National Geographic* 142 (1972): 783–802.

Tanner, Helen Hornbeck, ed. *Atlas of Great Lakes Indian History.* Norman, Okla., and London, 1987. Contains invaluable information about the Indians of the Illinois and adjacent regions.

Temple, Wayne C. *Indian Villages of the Illinois Country.* Springfield, Ill., 1958. Useful summary of what was known about the Illinois tribes in 1958.

Vogel, Virgil J. *Indian Place Names in Illinois.* Springfield, Ill., 1963. An excellent compendium of Indian names within the state, with many useful historical leads.

Winslow, Charles S., ed. *Indians of the Chicago Region.* Chicago, 1946. A short and popular book but a good compendium.

CHAPTER 3

Alvord, Clarence W. *The Illinois Country 1673–1818.* Chicago, 1922. Classic treatment by one of the great authorities.

Babson, Jane. "The Architecture of Early Illinois Forts." *Journal of the Illinois State Historical Society* 61 (1968): 9–40. Excellent survey with well-chosen plates.

Belting, Natalia. *Kaskaskia under the French Regime.* Urbana, Ill., 1948. Still the best summary of life at Kaskaskia under the French.

———. "When Illinois Was French." In *Illinois: Its History and Legacy,* ed. Roger D. Bridges and Rodney O. David. Saint Louis, 1984.

Boggess, A. C. *The Settlement of Illinois 1778–1830.* Chicago, 1908. Interesting treatment of the relationship between the new Anglo settlers and the French and Indians.

Brown, Margaret Kimball. "Fort Kaskaskia Evokes Memories of Conquest and Settlement." *Historic Illinois,* no. 3 (1979): 12–13.

Downer, Alan S. "Photo Reveals Probable Location of First Fort de Chartres." *Historic Illinois* 3, no. 2 (1980): 11.

Ekberg, Carl. *Colonial Sainte-Genevieve.* Saint Louis, 1985. A well-documented history of the town.

Fellows, Paul. "Fort Massac." *Historic Illinois,* no. 1 (1981): 1–13.

Franzwa, Gregory M. *The Story of Old Ste. Genevieve.* Saint Louis, 1967. Popular history of this fascinating town.

Gentilcore, R. Louis. "Vincennes and the French Settlement in the Old Northwest." *Annals of the Association of American Geographers* 67 (1957): 285–97.

Gumms, Bonnie. "Remnants of French Colonial Village Excavated at Cahokia." *Historic Illinois* 10, no. 3 (1987): 1–5.

Hart, John Fraser. *The Look of the Land.* Englewood Cliffs, N.J., 1975.

Imlay, Gilbert. *A Topographical Description of the Western Territory of North America.* London, 1797. Contains a series of chapters on different themes; the one by Thomas Hutchins on the Illinois country is very informative.

James, Alfred Procter, and Charles Morse Stotz. *Drums in the Forest.* Pittsburgh, 1958. Excellent analysis of various frontier fortifications.

McDermott, J. F. ed. *The French in the Mississippi Valley.* Urbana, Ill., 1965. Excellent chapters by Joseph Donnelly (the French between 1768 and 1778), Charles Peterson (French architecture), and Samuel Wilson (colonial fortifications).

———. *Old Cahokia.* Saint Louis, 1949. Still the best treatment of life at the village of Cahokia.

Marschner, F. J. *Land Use and Its Patterns in the United States.* Washington, 1959. Summary of landholding systems, followed by aerial photographs with commentary.

Moore, Evelyn. "The Cahokia Courthouse." *Historic Illinois,* no. 2 (1980): 1–3.

Norris, F. Terry. "Old Cahokia: An Eighteenth-Century Archaeological Site-Model." *Newsletter of the Center for French Colonial Studies* 2 (1984).

Orser, Charles E., and Theodore Karamanski. *Preliminary Archaeological Research at Fort Kaskaskia, Randolph County, Illinois.* Carbondale,

Ill., 1977. This substantial report disentangles the complicated historical and archaeological record of Fort Kaskaskia.

Peterson, Charles E. "Notes on Old Cahokia." *Journal of the Illinois State Historical Society* 42 (1949): 7–29, 193–258, 313–84.

Pittman, Philip. *The Present State of the European Settlements on the Mississippi.* Ed. J. F. McDermott. London, 1770; Memphis, 1976. Has a good plan and detailed descriptions of each of the French villages.

Price, Anna. "French Outpost on the Mississippi." *Historic Illinois*, no. 1 (1980): 1–4.

———. "The French Regime in Illinois." *Historic Illinois*, no. 3 (1982): 1–5; 5, no. 4 (1982): 1–6.

Tucker, Sara Jones. *Indian Villages of the Illinois Country: Part 1: Atlas.* Springfield, Ill., 1942. Reproductions of many of the maps needed to study the French presence in Illinois.

CHAPTER 4

Allen, John W. *It Happened in Southern Illinois.* Carbondale, Ill., 1968.

Barrett, Jay. *Evolution of the Ordinance of 1787.* New York, 1891. Detailed account of the emergence of the Northwest Ordinance.

Bergen, John. "Maps and Their Makers in Early Illinois: The Burr Map and the Peck-Messenger Map." *Western Illinois Regional Studies* 10 (1987): 5–31.

Billington, Ray. "The Historians of the Northwest Ordinance." *Journal of the Illinois State Historical Society* 40 (1947): 397–418. Masterly survey of the historiography of the ordinance.

Birkbeck, Morris. *Letters from Illinois.* Philadelphia, 1818.

———. *Notes on a Journey from the Coast of Virginia to the Territory of Illinois.* Philadelphia, 1817.

Black, Harry G. *Pictorial Americana: The National Road.* Hammond, Ind., 1984. Interesting details on local survivals.

Boewe, Charles. *Prairie Albion: An English Settlement in Pioneer Illinois.* Carbondale, Ill., 1962. Sympathetic account of the adventures of Birkbeck and Flower.

Boggess, A. C. *The Settlement of Illinois, 1778–1830.* Chicago, 1908. Meticulous account of the phases of immigration.

Buck, Solon. *Illinois in 1818.* Urbana, Ill., 1917. Interesting analysis of the state at its foundation.

Buley, R. Carlyle. *The Old Northwest: Pioneer Period, 1815–1840.* 2 vols. Indianapolis, 1950.

Cole, Harry Ellsworth. *Stagecoach and Tavern Tales of the Old Northwest.* Cleveland, 1930. Rather "antiquarian" in flavor, but has some telling resonances from the period, some of whose survivors Cole knew.

Hardin, Thomas. "The National Road in Illinois." *Journal of the Illinois State Historical Society* 60 (1967): 5–22.

Johnson, Hildegard Binder. *Order upon the Land.* New York, 1976. Analysis of the township-and-range system in the upper Mississippi country.

Lansden, John M. *A History of the City of Cairo, Illinois.* Chicago, 1910. Interesting account through local eyes.

Lawler, Lucille. *Gallatin County: Gateway to Illinois.* East Saint Louis, Ill., 1968.

McCall, Edith. *Conquering the Rivers.* Baton Rouge and London, 1984. Good account of the progress of inland navigation about 1840.

Miller, Keith L. "Planning, Proper Hygiene, and a Doctor: The Good Health of the English Settlement." *Journal of the Illinois State Historical Society* 71 (1978): 22–29. Fascinating account of the measures taken by Birkbeck and Flower to ensure the health of their settlers.

Moore, Evelyn. "Kaskaskia, Illinois' First Capital." *Historic Illinois*, no. 1 (1958): 8–9.

Pease, Theodore Calvin. *The Frontier State: 1818–1848.* Chicago, 1919.

Pooley, William Vipond. *The Settlement of Illinois from 1830 to 1850.* Madison, Wis., 1908. A good sequel to Boggess.

Quaife, Milo P. "The Significance of the Ordinance of 1787." *Journal of the Illinois State Historical Society* 30 (1938): 415–28.

Rathbun, Peter. "The Restoration of the Bank of Illinois at Shawneetown." *Historic Illinois*, no. 6 (1983): 6–11.

Stewart, George R. *U.S. 40: Cross-section of the U.S.A.* Boston, 1953; Westport, 1973.

Thomas, Benjamin P. *Lincoln's New Salem.* Springfield, Ill., 1934.

Woods, John. *Two Years' Residence on the English Prairie of Illinois.* Ed. Paul Angle. Chicago, 1968.

CHAPTER 5

Beecher, Edward. *Narrative of Riots at Alton: In Connection with the Death of Rev. Elijah P. Lovejoy.* Alton, Ill., 1838.

Broehl, Wayne G. *John Deere's Company.* New York, 1984.

Flagler, D. W. *A History of the Rock Island Arsenal.* Washington, 1877.

Historic Rock Island, Illinois. Rock Island, 1908.

Isaksson, Olov. *Bishop Hill, Ill.: A Utopia on the Prairie.* Stockholm, 1869.

Landrum, Carl A. *Historical Sketches of Quincy.* Quincy, Ill., ca. 1973.

Lewis, Henry. *The Valley of the Mississippi Illustrated.* Ed. Bertha Heilbron. Saint Paul, 1967. An edited version of *Das illustrierte Mississippithal.*

McManis, Douglas R. *The Initial Evaluation and Utilization of the Illinois Prairies, 1815–1840.* Chicago, 1964. An interesting monograph tending to analyze critically the old theories about the neglect of the prairie by the early settlers.

Meese, William A. *Early Rock Island.* Moline, Ill., 1905.

Mikkelsen, Michael A. *The Bishop Hill Colony.* Baltimore, 1892.

Moore, Evelyn R. "Eagle's Nest Artist Colony, 1898–1942." *Historic Illinois,* no. 2 (1984): 2–5.

Pease, Theodore Calvin. *The Frontier State, 1818–1848.* Chicago, 1919.

Peoria County, Illinois: History. Chicago, 1880.

Pooley, William Vipond. *The Settlement of Illinois from 1830 to 1859.* Madison, Wis., 1908.

Redmond, Pat H. *History of Quincy and Its Men of Mark.* Quincy, Ill., 1869.

Slattery, Thomas J. *An Illustrated History of the Rock Island Arsenal and Arsenal Island.* Rock Island, Ill., 1988. A full and recent summary of federal activities on Rock Island.

Smith, Laura Chase. *The Life of Philander Chase.* New York, 1903. His granddaughter's account of the life of the founder of Kenyon and Jubilee colleges.

Willis, John Randolph. *God's Frontiersmen: The Yale Band in Illinois.* Washington, 1979. Interesting account of Yankee influence in early Illinois.

graphical approach to the history of Chicago, with good aerial photographs.

Hatcher, Harlan. *The Great Lakes.* London and New York, 1944. A good popular summary by a well-known local historian.

Howe, Walter A. *Documentary History of the Illinois and Michigan Canal.* Springfield, Ill., 1956. Useful collection of original documents.

Knight, Robert, and Lucius Zeuch. *The Location of the Chicago Portage Route of the Seventeenth Century.* Chicago, 1928. Meticulous local history of the portage area.

Lamb, John. *A Corridor in Time: I. and M. Canal, 1836–1986.* Romeoville, Ill., 1987. A useful pamphlet by the leading local authority on the canal.

Mayer, Harold M., and Richard C. Wade. *Chicago: Growth of a Metropolis.* Chicago, 1973. A history that contains many unusual and interesting illustrations.

Noble, Vergil E. "History and Archaeology along the Illinois and Michigan Canal." *Historic Illinois* 9, no. 4 (1986): 10–15. A summary of recent work.

Peterson, Jacqueline. "Goodbye, Madore Beaubien: The Americanization of Early Chicago Society." *Chicago History* (summer 1980): 98–111.

Putnam, James W. *The Illinois and Michigan Canal.* Chicago, 1918. This remains the only full treatment of the canal's history.

Quaife, Milo P. *Lake Michigan.* Indianapolis, 1944. A good summary by one of the region's leading local historians.

Thornton, Nancy. "Saint James Church and Cemetery." *Historic Illinois* 8, no. 1 (1985): 5–6. An article arising out of the renewed interest in the canal area.

CHAPTER 6

Andreas, Alfred T. *History of Chicago from the Earliest Period to the Present Time.* 3 vols. Chicago, 1884–86. Still a great compendium of information.

Barry, James P. *Ships of the Great Lakes.* Berkeley, 1973. Offers a well-illustrated survey of this neglected subject.

Buley, R. Carlyle. *The Northwest: Pioneer Period, 1815–1840.* 2 vols. Indianapolis, 1950. Good sections on transport by land and water.

Conzen, Michael P., and Kay J. Carr. *The Illinois and Michigan Canal National Heritage Corridor: A Guide to Its History and Sources.* De Kalb, Ill., 1988. A bibliographic survey that ought to revive interest in the area.

Conzen, Michael P., and Melissa J. Morales. *Settling the Upper Illinois Valley.* Chicago, 1989.

Cutler, Irving. *Chicago: Metropolis of the Mid-continent.* Dubuque, Iowa, 1976. A rather geo-

CHAPTER 7

Bach, Ira, and Susan Wolfson. *A Guide to Chicago's Historic Suburbs.* Chicago, 1981. A good architectural guide, with some historical commentary, covering Lake, McHenry, Kane, Du Page, Will, and Cook counties.

Broehl, Wayne G. *John Deere's Company.* New York, 1984. A substantial but not triumphalist history of the great manufacturers of agricultural machinery.

Budd, Lillian. *Footsteps on the Tall Grass Prairie.* Lombard, Ill., 1977. This history of Lombard is one of the best of the local histories, which are the indispensable source for building up the larger picture of what was going on in northeastern Illinois.

Buley, R. Carlyle. *The Old Northwest: Pioneer Period, 1815–1840.* 2 vols. Indianapolis, 1950.

Chiles, James R. "Breaking Codes Was This Couple's Lifetime Career." *Smithsonian*, June 1987, 128–44. Fascinating summary of the work of the Friedmans and Riverbank Laboratories.

Cole, Harry Ellsworth. *Stagecoach and Tavern Tales of the Old Northwest*. Cleveland, 1930. A well-illustrated anecdotal work, full of charm.

Ehresmann, Julia M., ed. *Geneva, Illinois: A History of Its Times and Places*. Geneva, 1977. Another well-written local history, this time the work of a consortium of authors.

Ernst, Erik. "John A. Kennicott of the Grove." *Journal of the Illinois State Historical Society* 74 (1981): 109–18.

Hansen, Marcus Lee. *The Atlantic Migration 1607–1860*. Cambridge, 1940. Useful overview of the process of migration.

Kane County Development Department. *Rural Structures Survey: Preliminary Report*. Geneva, Ill., 1970.

Larson, Darlene, and Laura Hiebert. *The Fabyan Legacy*. Geneva, Ill., 1977. Good summary of the work of George and Nellie Fabyan.

McLear, Patrick. "The Galena and Chicago Union Railroad: A Symbol of Chicago's Economic Maturity." *Journal of the Illinois State Historical Society* 73 (1980): 17–26.

Nelson, C. Hal, ed. *Sinnissippi Saga*. Rockford, Ill., 1968. A history of Rockford and Winnebago County.

Quaife, Milo P. *Chicago's Highways Old and New*. Chicago, 1923. This masterly book has much information about the settlement of the northeast outside Chicago.

Stilgoe, John R. *Common Landscape of America, 1580 to 1845*. New Haven and London, 1982. This wide-ranging book, particularly strong on etymology and specific site analysis, allows the person interested in the history of Illinois to refer it to general trends in North America.

Swanson, Leslie. *Old Mills in the Midwest*. Moline, Ill., 1963. This short book has good maps and plates.

Thompson, Richard A., ed. *DuPage Roots*. Du Page County, Ill., 1985. This interesting local history, a collaborative work, consists of an introduction to the general history of the county, followed by short summaries of the history of each township.

Thornton, John A. *Reminiscences . . . of Early Days in Rockford*. N.p., 1891. A detailed and interesting account of early settler life.

Walters, William D., and Floyd Mansberger. "Early Mill Location in Northern Illinois." *Bulletin of the Illinois Geographical Society* 25, no. 2 (1983): 3–11.

Yoder, Paton. *Taverns and Travellers: Inns of the Early Midwest*. Bloomington, Ind., 1969. A more scholarly version of the book by Cole.

CHAPTER 8

Alexander, Edwin P. *Down at the Depot: American Railroad Stations from 1831 to 1920*. New York, 1970. A compendium of photographs, some of them illuminating.

Bach, Ira, and Susan Wolfson. *A Guide to Chicago's Train Stations, Past and Present*. Athens, Ohio, 1986. An illustrated listing of stations, many of the finest now destroyed.

Belcher, W. W. *The Economic Rivalry between St. Louis and Chicago, 1850–1880*. New York, 1947. Convincing explanation of the failure of Saint Louis to grow as rapidly as did Chicago.

Casey, Robert J., and W. A. S. Douglas. *Pioneer Railroad: The Story of the Chicago and Northwestern System*. New York, 1958.

Cooley, Thomas, ed. *The American Railway*. Chicago, 1889. Description of railroad operations, with many fine plates.

Gates, Paul Wallace. *The Illinois Central Railroad and Its Colonization Work*. Cambridge, 1934.

Kirkman, Marshall M. *Building and Repairing Railways*. New York and Chicago, 1904. Elementary operating manual with delightful plates.

McLear, Patrick. "The Galena and Chicago Union Railroad: A Symbol of Chicago's Economic Maturity." *Journal of the Illinois State Historical Society* 73 (1980): 17–26.

Meeks, Carroll V. *The Railroad Station: An Architectural History*. New Haven, 1956. Puts the stations of North America into their worldwide historical context, with excellent plates.

Nock, O. S. *Railways of the U.S.A.* New York, 1979. Summary of United States railroad operations by an expert English enthusiast.

Overton, Richard. *Burlington Route*. New York, 1965. History of the Chicago, Burlington, and Quincy from its beginnings in Aurora.

Pinkepank, Jerry A. "A Railroad's Railroad." *Trains*, September-October 1966. Original and fascinating history of the operations of the Belt Railway of Chicago.

Reps, John W. *The Making of Urban America*. Princeton, 1965. Contains a very interesting chapter on towns on the railroad.

Stilgoe, John R. *Metropolitan Corridor: Railroads and the American Scene*. New Haven and London, 1983. Concentrates on the neglected period between 1880 and 1930.

Stover, John. *History of the Illinois Central Railroad*. New York, 1975. Straightforward account, stressing the continuity of the routes served from the 1880s onwards.

Wade, Louise Carroll. *Chicago's Pride: The Stockyards, Packingtown, and Environs in the Nineteenth Century.* Urbana, Ill., 1987. Emphasizes the positive aspects of the meat-packing industry, slandered by Upton Sinclair's *The Jungle* of 1906.

CHAPTER 9

Abbott, Edith. *The Tenements of Chicago, 1908–1935.* Chicago, 1936.

Angle, Paul. *The Great Chicago Fire.* Chicago, 1946. A selection of firsthand accounts, illustrated with well-chosen plates.

Appleton, John B. *The Iron and Steel Industries of the Calumet District.* Urbana, Ill., 1927.

Bach, Ira. *Chicago on Foot: Walking Tours of Chicago's Architecture.* Chicago, 1973.

Buder, Stanley. *Pullman: An Experiment in Industrial Order and Community Planning, 1880–1930.* New York, 1967.

Cromie, Robert. *The Great Chicago Fire.* New York, 1958.

Cronon, William J. "To Be the Central City: Chicago, 1848–1857." *Chicago History* 10 (1981): 130–40. Short but perceptive account of Chicago's rise to economic supremacy in the Midwest.

Drury, John. *Old Chicago Houses.* Chicago, 1975.

Ericsson, Henry. *Sixty Years a Builder.* Chicago, 1942.

Jones, Peter d'A., and Melvin G. Holli. *Ethnic Chicago.* Grand Rapids, Mich., ca. 1981. Remarkable collection of essays on the ethnic communities of Chicago.

Lowe, David. *Lost Chicago.* Boston, 1975. Fascinating and distressing account of buildings destroyed around Chicago.

Swank, James. *History of the Manufacture of Iron.* Philadelphia, 1892.

The Tunnels and Water System of Chicago. Chicago, 1874.

CHAPTER 10

Appleton, John B. *The Iron and Steel Industries of the Calumet District.* Urbana, Ill., 1927. Interesting survey that brings out the remarkable advantages of the area for large-scale production.

Bach, Ira. "A Reconsideration of the 1909 'Plan of Chicago.'" *Journal of the Illinois State Historical Society* (spring/summer 1973): 132–41.

Bancroft, Hubert Howe. *The Book of the Fair.* New York, 1894; reprinted New York, n.d. A full and well-illustrated account of the various displays.

Block, Jean. *The Uses of Gothic.* Chicago, 1983. First published as an exhibition catalog, this well-illustrated book is full of interesting reflections about the choice of Gothic as a style for the university.

Burg, David. *Chicago's White City of 1893.* Lexington, Ky., 1976. A lively but scholarly account of the exposition, setting it into its political and cultural context.

Burnham, Daniel, and Edward H. Bennett. *Plan of Chicago.* Chicago, 1909; New York, 1970. A facsimile of the original edition of the plan, but one that gives a good idea of the excellence of its presentation.

Chicago Public Library. *CPL: The Historical Development of an Urban Library.* Chicago, 1985. This catalog of an exhibition held in the summer of 1985 gives a good idea of the emergence and growth of the library.

Condit, Carl W. *The Chicago School of Architecture . . . 1875–1925.* Chicago, 1964. A compendious account of building during this period.

Davies, Allen F. *American Heroine: The Life and Legend of Jane Addams.* New York, 1973.

Hines, Thomas. *Burnham of Chicago: Architect and Planner.* New York, 1974. A full and sympathetic account of Burnham's life and work.

Holt, Glen E. "Private Plans for Public Spaces." *Chicago History* (fall 1979): 173–84. Interesting analysis of the movement towards public parks in Chicago before Burnham.

Kantowicz, Edward R. *Corporation Sole: Cardinal Mundelein and Chicago Catholicism.* Notre Dame, Ind., 1983. A frank and opinionated account of the life and work of Cardinal Mundelein.

Kogan, Bernard. "Chicago's Pier." *Chicago History* (spring 1976): 28–38. Good summary of the history of Navy Pier.

Lane, George A. *Chicago Churches and Synagogues.* Chicago, 1981. An exceptionally well illustrated survey of ecclesiastical architecture.

Linn, James Weber. *Jane Addams: A Biography.* New York and London, 1935.

Roper, Laura Wood. *FLO: A Biography of Frederick Law Olmsted.* Baltimore and London, 1973.

Smith, Nina. "This Bleak Situation: The Founding of Fort Sheridan, Illinois." *Journal of the Illinois State Historical Society* 80 (1987): 13–22.

Stevenson, Elizabeth. *Park Maker: A Life of Frederick Law Olmsted.* New York and London, 1977.

Wille, Lois. *Forever Open, Clear, and Free: The Struggle for Chicago's Lakefront.* Chicago, 1972. A brisk and forceful account of the problems involved in retaining public access to the lakefront.

CHAPTER 11

Bouilly, Robert H., and Thomas J. Slattery. *Rock Island Arsenal: A Historical Tour Guide.* Rock Island, Ill., n.d. A recently published collection of commentated plates, giving an excellent idea of the development of the arsenal.

Conzen, Michael P., ed. *Time and Place in Joliet: Essays on the Geographical Evolution of the City.* Chicago, 1988. Some novel geographical perspectives on the development of a rather neglected city.

Flagler, D. W. *A History of the Rock Island Arsenal.* Washington, 1877.

Illinois Blue Book. The annual issues of the Blue Book contain statistics concerning population, production, and so forth.

Scamehorn, Howard L. *From Balloons to Jets, 1855–1955: A Century of Aeronautics in Illinois.* Chicago, 1957.

Tillinghast, B. F. *Rock Island Arsenal in Peace and War.* Chicago, 1898.

Waller, Robert A. "The Illinois Waterway from Conception to Completion." *Journal of the Illinois State Historical Society* 65 (1972): 125–41.

Weller, Allen S. *100 Years of Campus Architecture at the University of Illinois.* Urbana, Ill., 1968.

Yeater, Mary M. "The Hennepin Canal." *American Canals,* November 1976 to August 1978.

INDEX